ELECTRICAL WIRING

RESIDENTIAL
UTILITY
SERVICE AREAS

SIXTH EDITION

THOMAS S. COLVIN

**American Association for
Vocational Instructional Materials**
*220 Smithonia Road
Winterville, Georgia 30683*

The American Association for Vocational Instructional Materials (AAVIM) is a non-profit assiciation of American universitites, colleges, and divisions of vocational education. AAVIM's purpose is to prepare, publish, and distribute instructional materials for effective teaching and learning. Direction is given to the staff by representatives from the member states and agencies. AAVIM also works with teacher organizations, government agencies, and industry to provide accuracy and excellence in instructional materials.

AAVIM Staff

George W. Smith, Jr.	*Director*
James E. Wren	*Assistant Director*
Karen Seabaugh	*Office Manager*
Laura Ebbert	*Business Manager*
Rhonda Sanders	*Business Office*
Dean Roberts	*Shipping*
Margaret Ferruza	*Orders*
James A. Anderson	*Art Associate*
Suzanne Gilbert	*Typography*

Electrical Wiring/Sixth Edition Editorial Staff

George W. Smith, Jr.	*Editor*
James E. Wren	*Art Direction*
Suzanne Gilbert	*Typography*
James A. Anderson	*Production Art*

Sixth Edition

Copyright © 1993 by the American Association for Vocational Instructional Materials, 220 Smithonia Road, Winterville, Georgia 30683
Phone (706)742-5355

The information in this publication has been compiled from authoritative sources. Every effort has been made to attain accuracy; however, AAVIM and its associated individuals are not responsible for misapplication or misinterpretation of this information and cannot assume liability for the safety of persons using this publication.

An Equal Opportunity/Affirmative Action Institution

Printed in the United States of America

ISBN 0–89606–302–X

Author

Thomas S. Colvin is credited with the writing of the first edition of this publication. He is also responsible for writing the updated information in the second, third, fourth, and fifth editions. Mr. Colvin served as Research and Development Specialist of AAVIM at the time of the initial writing. He has since retired.

Electrical Wiring Editorial Staff–Previous Editions

First and Second Editions
J. Howard Turner, former Editor and Coordinator, AAVIM, was the editor of the first edition. All original art and graphics were the work of James E. Wren, Art Director, AAVIM; George W. Smith, Jr., Production Coordinator, AAVIM; and Thomas B. Brown, former artist with AAVIM.

Third Edition
Richard M. Hylton, former Executive Director, AAVIM, served as editor for this edition. Updated and new illustrations and graphic changes were the work of James E. Wren, Art Director, AAVIM; and Donna D. Pritchett, former artist with AAVIM.

Fourth, Fifth, and Sixth Editions
George W. Smith, Jr. acted as editor for the revised copy, while James E. Wren was responsible for new and updated illustrations. Typesetting of new and revised copy was the work of Suzanne Gilbert.

Consultants

Former Editions
Dr. James M. Allison acted as a consultant on the first, second, third, fourth and fifth editions of this publication. Dr. Allison is currently Professor of Electrical Engineering in the Division of Biological and Agricultural Engineering at the University of Georgia where he teaches courses dealing with electricity.

Sixth Edition
M. Ralph Duncan acted as consultant for the revisions in the sixth edition of this publication. Mr. Duncan is Senior Methods and Training Specialist, Georgia Power Company, Atlanta, GA.

LaVerne E. Stetson, P. E., Agricultural Engineer, USDA-ARS, Lincoln, Nebraska, reviewed the text and provided technical assistance and suggestions.

Contents

Preface

This manual will be useful for those who are preparing to become electricians and for do-it-yourselfers who may be installing electrical wiring. All procedures are fully explained and illustrated. If proceeding on your own, please pay close attention to all safety procedures outlined in this book. All procedures recommended in this manual are based on provisions of the National Electrical Code.®*. This sixth edition has been revised to include all pertinent changes in the 1993 National Electrical Code® which remains in effect until 1996. Wiring installed according to procedures given should provide for an installation that meets Code requirements. Consult the National Electrical Code® for any needed additional information. Procedures are given for complete installation of the wiring systems for residences, utility buildings and service areas. In some localities, wiring by unlicensed persons may be prohibited. Check your local and state laws.

National Electrical Code® and NEC® are registered trademarks of the National Fire Protection Association, Inc., Quincy, MA 02269.

OBJECTIVES:

Upon successful completion of this course, you will be able to perform the following:

— Read and interpret wiring plans.
— Locate and mark routes for small appliance, general purpose and individual circuits.
— Install device (switch) boxes and outlet boxes.
— Install 120-volt, 120/240-volt and 240-volt circuits.
— Ground the electrical system and equipment.
— Connect receptacles, switches and fixtures for each circuit.
— Determine type and size of service entrance equipment to install.
— Install service entrance equipment using cable or conduit with overhead or underground conductors.
— Install ground fault circuit interrupter.
— Install conduit.
— Estimate wiring costs.
— Install wiring for agricultural and other utility buildings.

NOTE: SI (metric) conversions are given in approximate values. SI terms are given first and U. S. Customary terms in parentheses.

Other instructional material
dealing with electricity
available from AAVIM:

—**Understanding Electricity and Electrical Terms.**
—**Home Electrical Repair.**
—**Electric Energy—Utilization, Generation, Transmission, Distribution, Conservation.**
—**How Electric Motors Start and Run.**
—**Electric Motors—Selection, Protection, Drives.**

Available for classroom assistance for this publication are a teacher guide, student workbook and an audiovisual. Contact AAVIM for current prices and ordering information.

Introduction

A wiring system that is well planned and installed correctly will supply power for all of the owner's electrical needs (Figure 1). It will also provide for additions that might be required in the future. To be assured that wiring is installed correctly, most communities have adopted a *wiring code* that governs the installation of all types of electric wiring. Local codes may have requirements that are more strict than the National Electrical Code®.

There are two organizations with which you should be familiar:

—**National Fire Protection Association (NFPA).**
—**Underwriters Laboratories (UL).**

The **National Fire Protection Association,** under the auspices of the American National Standards Institute (ANSI), sponsors the **National Electrical Code®** which provides minimum standards for the safe elec-

FIGURE 1. **A well planned housewiring system supplies power for all electrical needs.**

FIGURE 2. The National Electrical Code® gives requirements for safe installation of electrical wiring.

trical wiring installations (Figure 2). The National Electrical Code® (NFP 70) is abbreviated "NEC" but is usually referred to as the "Code." When used in this manual, the Code means the National Electrical Code®. A condensed version of the Code (NFP 70A) which covers wiring procedures for one- or two-family dwellings is also available from NFPA.

Code tables included in the manual are copyrighted by the NFPA and are reprinted by special permission of the NFPA.

Throughout this publication, reference to particular sections of the code are indicated by a black box within the illustrations.

A revised version of the Code is issued every three years. Copies may be purchased by writing the "National Fire Protection Association," Batterymarch Park, Quincy, MA 02269.

The purpose of the Code is to assure safe installation of electric wiring for the protection of life and property. It does not have the force of law unless it is adopted by the governing bodies of the state, county, or local body where it will be applied. Local authorities may include additional requirements.

The National Electrical Code® is referred to or quoted in the manual in many places and interpretation may sometimes be implied. The reader should understand that any interpretation of the Code, implied or otherwise, is the interpretation of AAVIM and is not official. AAVIM cannot accept liability from the use of the book. However, the book has been reviewed by individuals from throughout the United States and Canada in an effort to make the information correct and understandable. Among those who reviewed the book are members of the National Electrical Code Committee, electrical contractors, electrical engineers and elec-

trical professors, electrical inspectors, and vocational instructors from high schools, vocational-technical schools and junior colleges.

You should have copies of both the Code and the condensed version and know how to use them. Keep up-to-date copies on hand.

The **Underwriters Laboratory** (UL) tests electrical wiring materials. The purpose of the tests is to determine if the products meet minimum standards for safety and quality which are established either by the laboratory, or by ANSI. You should look for the UL tag or stamp when buying wiring materials (Figure 3). This is your assurance that the products meet **minimum** safety standards. It does not mean that all items that are UL listed are the same quality. Materials made by several manufacturers may all carry the UL stamp. But one product may last much longer in service than the other. Any material that does not carry the UL tag or stamp may not be safe. It will not be approved for use by most electrical inspectors.

FIGURE 3. Underwriters Laboratory stamp or tag assures only that the product meets minimum safety standards.

Most localities require the electrician or owner to obtain a **permit** before starting electrical work. The permit is authorization to begin the wiring installation. One or more inspections are usually made by the local inspector. One may be made at the "rough-in" stage and another upon completion of the installation. The **local inspector** is the final authority in interpreting local codes and the National Electrical Code®.

Procedures for electrical wiring are given under the following headings:

    I. Reading and Interpreting Wiring Plans.

    II. Installing Wiring.

    III. Connecting Circuits.

    IV. Installing Service Entrance Equipment.

    V. Installing Metallic Conduit.

    VI. Estimating Wiring Costs.

    VII. Wiring Applications for Rural Areas.

I. Reading and Interpreting Wiring Plans

The purpose of this section is to provide basic understanding of wiring plans. Beginning with simple circuits, you will follow step-by-step the development of a house wiring plan.

On most house wiring jobs, the electrician works from the wiring plan shown on the floor plan of the house (Figure 4). **Symbols** are used to represent ceiling outlets, switches, convenience outlets (receptacles) and electrical equipment (Figure 5). Three of the most common symbols are the ceiling outlet, the switch and the duplex receptacle (Figure 6).

Upon successful completion of this section, you will be able to **identify electrical symbols** and **trace the route of each type of circuit** in a **house plan.**

Reading and interpreting wiring plans are discussed under the following headings:
A. Understanding Circuits.
B. Identifying Small Appliance Circuits.
C. Identifying General Purpose Circuits.
D. Identifying Individual Circuits.
E. Identifying Structural Parts of a House.

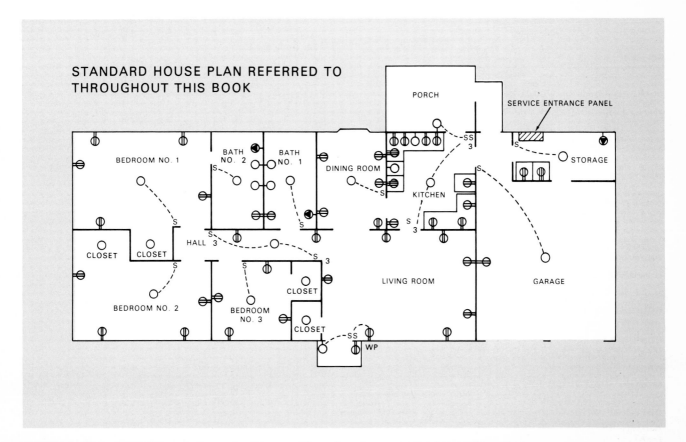

FIGURE 4. The electrician usually works from a wiring plan shown on the house floor plan.

COMMON ELECTRICAL SYMBOLS

Symbol	Description	Symbol	Description
O	CEILING OUTLET	▨	SERVICE ENTRANCE PANEL
—O	WALL OUTLET	S	SINGLE POLE SWITCH
O$_{PS}$	CEILING OUTLET WITH PULL SWITCH	S$_2$	DOUBLE POLE SWITCH
—O$_{PS}$	WALL OUTLET WITH PULL SWITCH	S$_3$	3-WAY SWITCH
⊖	DUPLEX CONVENIENCE OUTLET	S$_4$	4-WAY SWITCH
⊖$_{WP}$	WEATHERPROOF CONVENIENCE OUTLET	S$_P$	SWITCH WITH PILOT LIGHT
⊖$_{1,3}$	CONVENIENCE OUTLET 1=SINGLE 3=TRIPLE	▪	PUSH BUTTON
⊖$_R$	RANGE OUTLET	CH◦	BELL OR CHIMES
⊖$_S$	CONVENIENCE OUTLET WITH SWITCH	◄	TELEPHONE
⊖$_D$	DRYER OUTLET	TV	TELEVISION OUTLET
⊖	SPLIT WIRED DUPLEX OUTLET	S---	SWITCH WIRING
▲	SPECIAL PURPOSE OUTLET	▭	FLUORESCENT CEILING FIXTURE
D	ELECTRIC DOOR OPENER	◯◯	FLUORESCENT WALL FIXTURE

FIGURE 5. Examples of symbols used to represent electrical outlets, switches and equipment.

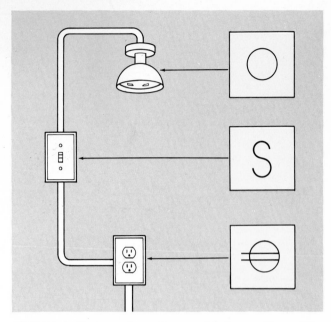

FIGURE 6. The most common symbols are the ceiling outlet, the switch and the duplex receptacle.

A. Understanding Circuits

Understanding circuits is discussed as follows:
1. Types of Electrical Circuits.
2. Current Carrying Capacity (Ampacity) of Electrical Circuits.

1. TYPES OF ELECTRICAL CIRCUITS

An electrical circuit may be defined as two or more conductors (wires) through which current flows from a source to one or more outlets.

To help you understand what is meant by the **source,** the sketch in Figure 7 shows the wiring from the transformer of the power supplier to the service equipment. From the transformer, the current flows through the meter to the **service entrance panel** (SEP) inside the house. This panel is considered the source for every circuit in the system. From the service entrance panel, the circuits branch out to all parts of the house. In fact, in residential wiring, most circuits that begin at the service entrance panel are known as **branch circuits.**

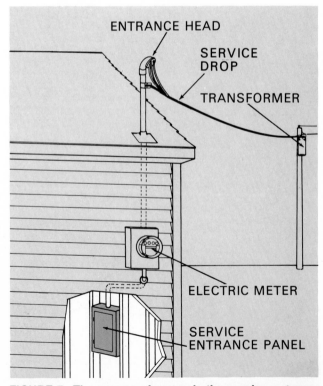

FIGURE 7. The source of power is the service entrance panel (SEP) which receives power from the supplier lines.

10

To have a circuit of any kind, you must have two or more conductors starting at the SEP and extending to one or more outlets. A **conductor** is any type of material that is used to carry electrical current. In most instances in this book, they are called "wires," but the terms "wire" and "conductor" are used interchangeably at times. Most circuits are made up of two or more insulated wires bound together in a "cable."

Types of electrical circuits are discussed as follows:

a. Branch Circuits.
b. Feeder Circuits.

a. Branch Circuits

A **branch circuit** is the circuit between the last fuse or circuit breaker and the outlet(s). Most circuits installed in homes are branch circuits. They run from the fuse or circuit breaker in the SEP to one or more outlets, such as a light or receptacle (Figure 8).

Having several circuits in a system makes it possible to use smaller and less expensive wire. One reason for this is to **keep costs down**. If only one circuit is used, larger and more expensive wire and materials are required.

There are three common branch circuits. They are:

—**Small appliance.**
—**General purpose.**
—**Individual equipment.**

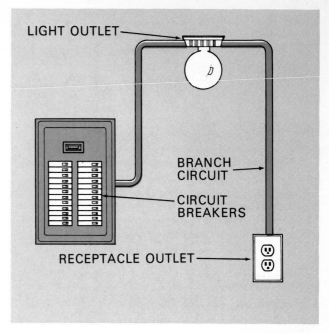

FIGURE 8. A branch circuit starts at the fuse or circuit breaker in the SEP and runs to one or more outlets.

b. Feeder Circuits

A feeder circuit is the circuit between the SEP and the sub-panel. Such a panel is normally used with smaller circuits. The sub-panel may be main lugs only (MOL). A feeder may also originate at a generator or battery. In larger homes or structures, a feeder circuit is sometimes required (Figure 9). In larger homes a feeder with a sub-panel is often used to elevate a potential voltage drop problem caused by long circuits. In this

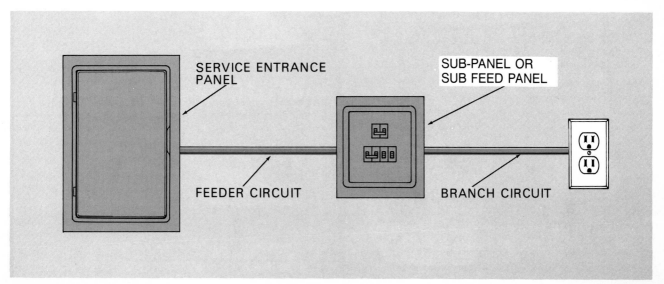

FIGURE 9. A feeder circuit extends between the SEP and a sub-panel or sub-feed panel protecting a branch circuit.

application a circuit can originate in a sub-panel branch circuit, this significantly reducing the branch circuit length.

In new installations, the authority having jurisdiction will most often require an outside disconnecting means to disconnect the service, which will result in a feeder cable serving a sub-panel board located centrally in the home. *NEC*® Code Section 230-70 (a) requires that the service disconnecting means be located either outside or immediately inside, adjacent to the point of entrance. This code section is now enforced in virtually all jurisdictions. This requirement has been a part of the *NEC*® for many code cycles, but only recently has the necessity for enforcement been recognized.

2. CURRENT CARRYING CAPACITY (AMPACITY) OF ELECTRICAL CIRCUITS

The **total amount of current required** in all parts of the system is known as the **electrical load**. The term is usually shortened and referred to as the "**load**." Different kinds of **circuits** are needed to carry the load in each room or combination of rooms.

The wiring in each circuit must be protected by an overcurrent device, either a **fuse** or a **circuit breaker**, sized to protect the circuit wiring (Figure 10). If the circuit becomes overloaded, or a short circuit develops, the fuse or circuit breaker will open to stop the flow of current. The circuit breaker is normally used in newer installations.

In addition to keeping costs down, another reason for providing several branch circuits is for **safe operation** of the system. Dividing the current flow among several circuits helps prevent overloading of any single circuit. By dividing load requirements into several circuits, a fuse or circuit breaker can be provided for each circuit (Figure 11). If trouble develops on one circuit, only that circuit will be out of operation when the fuse blows or circuit breaker trips. The load should be evenly divided among the branch circuits within the SEP.

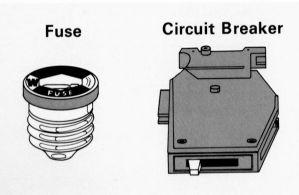

FIGURE 10. Each circuit must be protected by either a fuse or a circuit breaker.

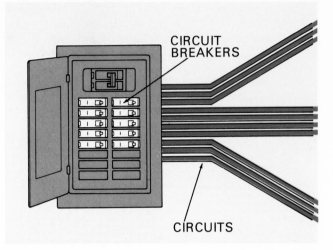

FIGURE 11. By dividing load requirements into several circuits, a fuse or circuit breaker can be provided for each circuit.

A large diameter water pipe can carry more water than a small pipe. In a similar way, large diameter wire will allow more electrical current to flow than a small wire (Figure 12). The load-carrying capacity of wire is measured in **amperes**. The load-carrying capacity of the wire stated in amperes is **ampacity**. The code defines ampacity as "the current in amperes a conductor can carry continuously under conditions of use without exceeding its temperature rating." As you would expect, the large diameter wire costs more than small diameter wire. Therefore, to keep costs down, the electrician chooses wire of the correct ampacity for each circuit. For example, the circuit for an electric range in the kitchen must carry one of the larger loads used in the house. Therefore, the circuit for the range

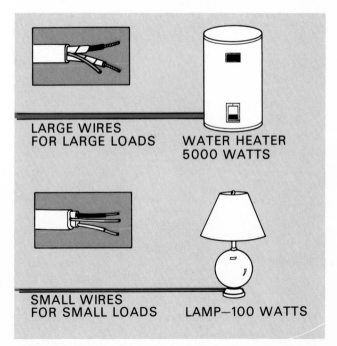

FIGURE 12. A large diameter wire will carry a larger current load than a small wire.

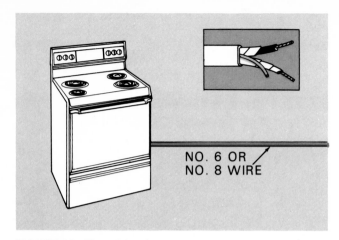

FIGURE 13. The wiring for a range must be large gage to provide the ampacity needed.

must be wired with large gage wire to provide the ampacity required (Figure 13). When using American Wire Gage (AWG) ratings, the smaller the number, the larger the wire. Thus, No. 8 AWG has a larger diameter than No. 10 AWG but smaller than No. 6 AWG (Figure 14). All numbers referring to wire size indicate the American Wire Gage size.

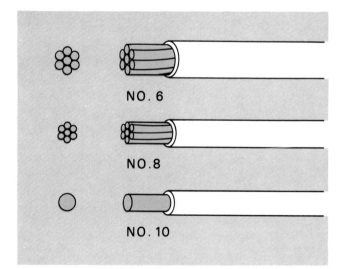

FIGURE 14. The smaller the AWG number, the larger the wire.

General purpose circuits for lighting carry the smallest loads in the house. The Code calls them lighting circuits but to call them general purpose circuits describes them better. They do supply lighting, but vacuum cleaners, radios, stereos, televisions and many other items are also connected to them. Number 12- or 14-AWG is used for general purpose circuits (Figure 15).

The **small appliance** circuit carries a larger load than the general purpose circuit but a much smaller load than the range circuit. The Code requires a minimum of two small appliance circuits. Each circuit is rated at 20 amperes.

Fuses and circuit breakers are rated in amperes. The rating used depends on the ampacity of the circuit it protects. Wire size and overcurrent protection for residential circuits are given in Table 1.

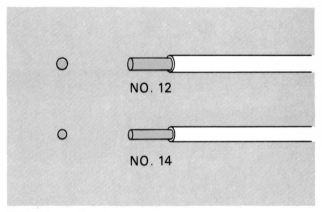

FIGURE 15. The Code allows the use of either No. 12 or No. 14 AWG for general purpose circuits.

TABLE I. WIRE SIZE AND OVERCURRENT PROTECTION FOR RESIDENTIAL CIRCUITS

Kind of Circuit	AWG Wire Size* (copper)	Maximum Overcurrent Protection (amperes)
General Purpose	14	15
	12	20
Small Appliance	12	20
Individual Equipment	10 or larger	30 or larger

*The larger the wire, the smaller the AWG number.

B. Identifying Small Appliance Circuits

Small appliances include **mixers**, **blenders**, **toasters**, **coffee makers**, **waffle irons**, **fry pans**, **refrigerators** and many others.

The Code requires that two or more **20-ampere small appliance circuits** be provided for receptacle outlets, in the kitchen, pantry, breakfast room and dining room. Code Section 210-52 (b) requires that the wire size for the small appliance circuit must be at least No. 12 (Figure 16). It is protected by a 20-ampere fuse or circuit breaker.

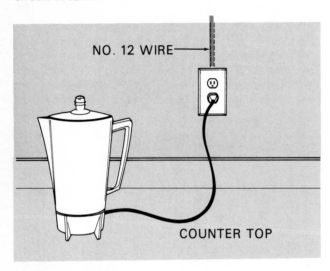

FIGURE 16. Wire size for small appliance circuits must be at least No. 12 AWG.

The Code also requires that a separate 20-ampere circuit be provided to supply the laundry receptacles. Code Section 210-52 (f) says that no other outlets can be connected to it.

At least two small appliance circuits must serve the **countertop kitchen receptacles** but they may also extend into the other rooms named above to serve receptacle outlets only. If more than two small appliance circuits are installed, the additional circuits may serve receptacle outlets in the other rooms specified above, if desired.

No lighting or other outlets are permitted on small appliance circuits, with five exceptions. According to Code Section 210-52 (b) Ex. 1 and 2, a clock outlet that supports and supplies power to a clock may be connected to a small appliance circuit and to outdoor receptacle outlets. In the new 1993 *NEC®*, it is permissible to supply electric ignition and timers for gas ranges, ovens, or countertop cooking units. Also, certain motor loads are now acceptable (Code section 210-52 (b) (Figure 17).

Even though refrigerators often require less current than many small appliances, they are required to be installed on a dedicated circuit.

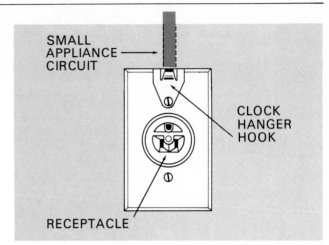

FIGURE 17. A clock outlet may be connected to a small appliance circuit.

Small appliance outlets are easily identified on the plan. No lighting outlets are included. All convenience outlets (receptacles) indicated for the kitchen, pantry, breakfast room and dining room make up the small appliance circuits (Figure 18) (Code Section 210-52 [b]).

In addition to required small appliance circuits, switched receptacles are now permitted for lamps in kitchens, dining rooms, breakfast rooms or pantry, provided they are supplied by a general purpose circuit rather than one of the 20-ampere small appliance circuits (Code Section 210-52 [b] Ex. 3).

The 1993 Code Section 210-52 (b) Ex. 4 allows a circuit supplying only motor loads while Code Section 210-52 (b) Ex. 5 allows circuits for electronic ignition systems for gas ranges, ovens, etc.

The parts of a small appliance circuit are shown in Figure 19. The circuit begins at a circuit breaker in the service entrance panel and continues through a stud wall to the kitchen counter where receptacles are installed.

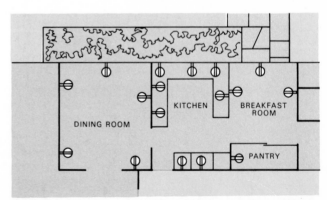

FIGURE 18. Small appliance circuits consist of all convenience outlets (receptacles) in the kitchen, pantry, breakfast room and dining room.

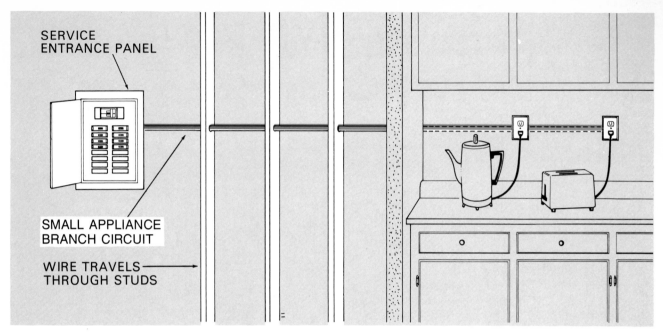

FIGURE 19. A small appliance circuit begins at the SEP and runs through studs to kitchen receptacles.

C. Identifying General Purpose Circuits

General purpose circuits make up the largest part of the wiring system for a home. Included are all ceiling and wall mounted lighting outlets and most receptacles (Figure 20). General purpose circuits are protected by either a 15- or 20-ampere fuse or circuit breaker, depending on whether No. 12 or No. 14 wire

FIGURE 20. General purpose circuits such as ceiling lights and receptacles in a bedroom make up the largest part of the wiring system for a home.

is used. Either a 15- or 20-ampere fuse or circuit breaker may be used with No. 12 wire. But No. 14 wire must be protected by a 15-ampere or smaller fuse or circuit breaker.

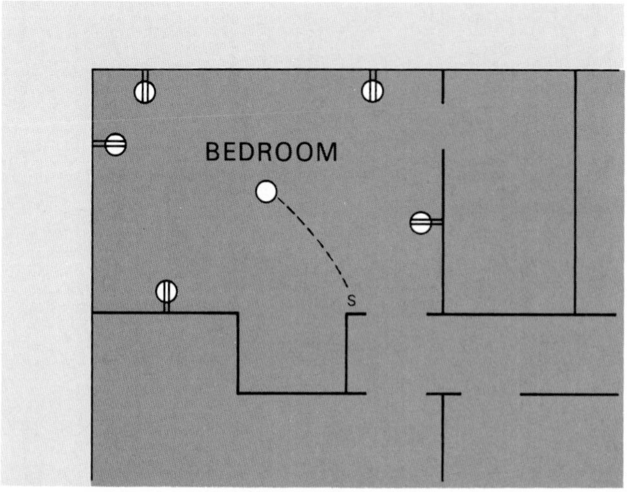

Identifying general purpose circuits is discussed as follows:

1. Determining Number of General Purpose Circuits.
2. Dividing Wiring into General Purpose Circuits.

1. DETERMINING NUMBER OF GENERAL PURPOSE CIRCUITS

How do you decide the number of general purpose circuits you will need for a house? The wiring plan shows the location of each outlet in the house. However, most plans do not indicate which outlets will be combined to make up a circuit. The electrician must decide how the circuits will be grouped to include all outlets called for on the wiring plan.

Code Section 220-2 provides detailed procedures for computing the required number of circuits. Instructions in the use of those procedures are given in Section IV, "Installing Service Entrance Equipment." The following procedures provide a less complicated method that is fairly accurate:

Most authorities agree that there should be a **minimum** of **one** general purpose circuit for every 46.5 square meters (500 square feet) of floor space. On this basis, a house containing 185.8 square meters (2,000 square feet) would require four general purpose circuits (2000 ÷ 500 = 4) (Figure 4). Most house plans will show the number of square meters (or square feet) in the house. If not, it can easily be found by multiplying the outside dimensions, leaving out open porches and garages. However, you should include any area which might be developed for future use, such as an unfinished attic or basement. If you have such an area, you should run one or more separate circuits to it. Provide at least one outlet if the area is used for storage or contains equipment requiring servicing, for example, a furnace. When the area is finished at a future date, the wiring can be extended to run additional outlets without danger of overloading.

An example of the above procedures is as follows:

1. *Find the number of square meters (or square feet) in the house to be wired by multiplying the length × width.*

    A house 12.1m x 15.2 m (40' x 50') contains 183.9 square meters (2000 square feet.)

    A two-story house 8.5 m × 12.2 m = 103.7 square meters (28' × 40' = 1120 square feet), multiplied by 2 for both floors = 207.4 square meters (2240 square feet).

2. *Divide floor area by 46.5 square meters (500 square feet) per circuit.*

    185.8 ÷ 46.5 (2000 ÷ 500) = 4 general purpose circuits (Figure 4), or 207.4 ÷ 46.5 (2240 ÷ 500) = 5 circuits (Always go to next highest number of circuits if division shows a fraction.)

It should be noted that demands for energy may often require additional circuits beyond those called for in a square foot calculation if you are making provisions for future expansion capacity.

2. DIVIDING WIRING INTO GENERAL PURPOSE CIRCUITS

After you have calculated the number of general purpose circuits required, your next step is to divide the outlets into the correct number of circuits. The total load, as nearly as practicable, must be evenly proportioned (balanced) among the number of circuits calculated.

When laying out general purpose circuits, you might not have the same number of outlets on each circuit. Some will have more than the average number of outlets computed from Table II. Some will have a smaller number. To avoid overloading, do not exceed 10 outlets per 15-ampere circuit or 13 per 20-ampere

general purpose circuit. When grouping outlets to make up a circuit, keep in mind the combinations that will require the least amout of wire to connect. Remember, every circuit may begin at the SEP or sub-panel.

Some rooms may have part of the outlets on one circuit and part on another. This is necessary to balance the load properly, and is desirable. With practice, you will soon be able to lay out the lighting circuits in a way that is both economical and serviceable.

Architects and electrical contractors sometimes use a system of lines drawn on house plans to indicate the exact route for each circuit. Some plans also indicate the number and size of wires for each circuit.

One procedure that may be used to balance the load for general purpose circuits is as follows:

1. *Count the number of outlets shown on the wiring plan and divide by the number of circuits required.*

    For this problem, assume 15-ampere circuits. Include in your count all ceiling and wall lights and all convenience outlets except **small appliance** and **individual** outlets. Individual outlets include those for the range, water heater, clothes dryer, and similar appliances.

    A table may be helpful that shows the following:

TABLE II. NUMBER OF OUTLETS IN GENERAL PURPOSE CIRCUITS
(Figures shown are taken from the house in Figure 4)

Location		Ceiling and Wall Lighting Outlets	Convenience Outlets*
Bedrooms & Closets	No. 1	2	5
	No. 2	2	5
	No. 3	2	4
Hall & Entry		3	5
Baths	No. 1	3	1
	No. 2	3	1
Living Room		1	6
Dining Room		1	–
Kitchen		3	–
Laundry or Utility Room		1	1
Outdoor Receptacles, Lights and Garage		3	3
Total Outlets in Each Column		**24**	**31**
Total Outlets in Both Columns		**55**	

*Note that convenience outlets on circuits for small appliances and for individual circuits are not included in dining room, kitchen and laundry room.

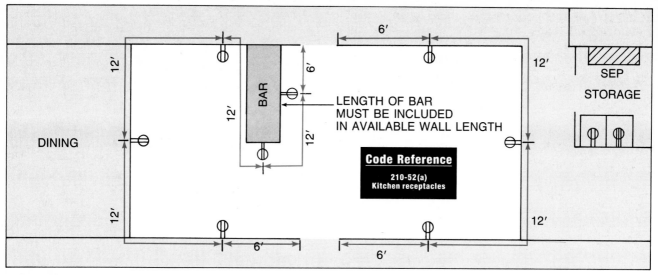

FIGURE 20A. Receptacle outlets are required to be positioned at intervals of 6 feet or less along an unbroken floor line. This line includes any wall space 2 feet wide or wider.

2. *Total the two columns and add them together.*

3. *Divide by the number of circuits called for to find the average number of outlets per circuit.*

If the average is more than 10 outlets per circuit, it is a good practice to add one or more circuits.

In Table II, the total number of general purpose outlets is 55. Divide by 4 (number of circuits computed on a square meter or square foot basis) and you will find that the average number of outlets per circuit is 13.7. As suggested above, if the average number of outlets per circuit is more than 10, another circuit should be added. By doing so, the number of general purpose circuits is now five for the house in Table II. Divide 55 outlets by 5 circuits (55 ÷ 5 = 11) and the average number of outlets per circuit is still more than ten. A sixth circuit must be added to provide enough capaci-

ty for the required number of outlets and to allow some extra capacity for future expansion. It is less expensive in the long run to allow some extra capacity in new homes than to add on later.

To meet Code requirement 210-52 (a), the wiring plan must show receptacle outlets installed so that no point along an unbroken wall line measures greater than 1.83 m (6 feet) between outlets including any wall space 610 mm (2 feet) or more in width (figure 20A). Code Section 210-52 (c) requires that kitchen countertop receptacles be installed at each counter space wider than 305 mm (12 inches). Kitchen countertop receptacles must be installed so that no distance measured between outlets will be greater than 610 mm (24 inches) (Figure 20B). In addition, a 125 volt single phase 15 or 20 amp receptacle must be installed to be convenient for the servicing of heating and air conditioning in crawl spaces and attics, according to Code Section 210-63.

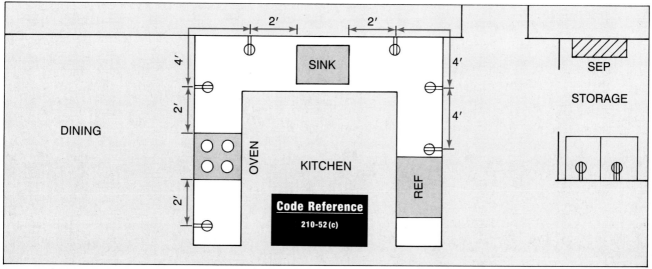

FIGURE 20B. Kitchen countertop receptacles at any wall counter space 12 inches wide or wider, must be positioned so that there is a outlet for every 4 linear feet (or fraction thereof) of countertop length.

At least one lighting outlet controlled by a light switch must be located at the entrance of the attic, basement or crawl space if the space is used for storage or equipment needing servicing (Code Section 210-70 [a]).

In addition, a wall switch controlled lighting outlet must be installed near heating and air conditioning equipment (Code Section 210-70 [c]).

All single family dwellings should have outdoor outlets in the front and back of the house (210-52 [e]).

D. Identifying Individual Circuits

An **individual** branch circuit is one that is wired directly from the entrance panel to one outlet or directly connected to only one appliance or item of electrical equipment (Figure 21). It is installed for items that require larger amounts of power than small appliance and lighting loads. For some individual circuits, larger gage wire and larger ampacity circuit breakers or fuses are required. Also, some appliances and equipment on individual circuits must be wired for 240 volts, or a combination of 120/240 volts.

As a general rule, the following are wired as separate or individual circuits:

- **Range, self contained.**
- **Range, built-in oven, separate range top.**
- **Range, built-in oven double oven; separate range top.**
- **Central electric heat.**
- **Clothes dryer.**
- **Water heater.**
- **Air conditioner.**
- **Dishwasher.**
- **Food freezer.**
- **Garbage disposer.**
- **Trash compactor.**
- **Fixed bathroom heater.**
- **Furnace motor and controls.**
- **Motors ½ hp and over.**
- **Washing machine circuit.**

Although not required by the Code, some power companies now recommend individual 20-ampere circuits for bathrooms. One reason is that many hand-held hair dryers are rated at 1500 watts, which requires 13 amperes. A small additional load would overload a 15-ampere circuit.

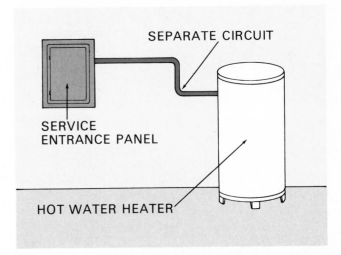

FIGURE 21. An individual circuit has separate wiring from the service entrance panel to the outlet for only one appliance or item of electrical equipment, such as the water heater.

18

E. Identifying Structural Parts of a House

House plans contain many terms that refer to parts of the house. You will become familiar with most of these as you gain experience. However, there are a few basic structural parts you should be able to identify from the start. Knowing the names of the parts will help you understand the location of electrical circuits more clearly.

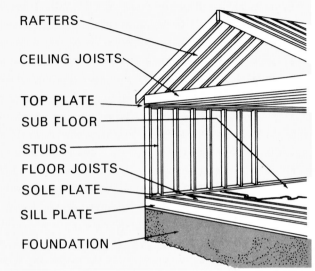

RAFTERS
CEILING JOISTS
TOP PLATE
SUB FLOOR
STUDS
FLOOR JOISTS
SOLE PLATE
SILL PLATE
FOUNDATION

FIGURE 22A. Much of the wiring system is attached to studs, floor and ceiling joists and top plates and sole plates.

On new work, you will run wiring and attach devices to the open framing (Figure 22). The horizontal members are **ceiling joists** and **floor joists.** These are usually 5.1 cm × 15.2 cm (2″ × 6″), 5.1 cm × 20.3 cm (2″ × 8″) or 5.1 cm × 25.4 cm (2″ × 10″) lumber.

The same structure has a **top plate** usually made up of two 5.1 cm × 10.2 cm (2″ × 4″'s) and a **sole plate** made of one 5.1 cm × 10.2 cm (2″ × 4″) (Figure 22). The lower plate is often referred to as the **sole.**

The uprights are the wall studs, usually constructed of 5.1 cm × 10.2 cm (2″ × 4″'s). Studs and plates in outside walls may be 5.1 cm × 15.2 cm (2″ × 6″'s) to allow for more insulation. Firestops, 5.1 cm × 10.2 cm (2″ × 4″) cross members, may or may not be included. Much of the wiring and many electrical devices are attached to the studs.

The **subfloor** is the first layer of flooring material over the floor joists (Figure 22). Several kinds of material may be used for the subfloor. They include 1.2 m × 2.4 m (4′ × 8′) sheets of plywood and 1.2 m × 2.4 m (4′ × 8′) sheets of particle board (pressed wood chips). Lumber in 2.5 cm × 15.2 cm (1″ × 6″) dimensions may also be used.

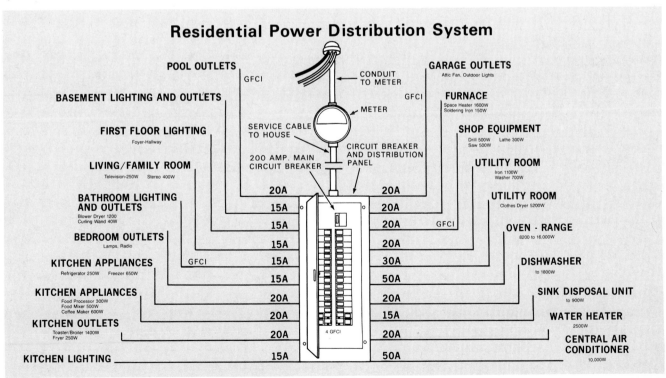

Residential Power Distribution System

POOL OUTLETS — GFCI
BASEMENT LIGHTING AND OUTLETS
FIRST FLOOR LIGHTING — Foyer-Hallway
LIVING/FAMILY ROOM — Television-250W Stereo 400W
BATHROOM LIGHTING AND OUTLETS — Blower Dryer 1200 Curling Wand 40W
BEDROOM OUTLETS — Lamps, Radio
KITCHEN APPLIANCES — Refrigerator 250W Freezer 650W — GFCI
KITCHEN APPLIANCES — Food Processor 300W Food Mixer 500W Coffee Maker 600W
KITCHEN OUTLETS — Toaster/Broiler 1400W Fryer 250W
KITCHEN LIGHTING

CONDUIT TO METER
SERVICE CABLE TO HOUSE
METER
200 AMP. MAIN CIRCUIT BREAKER
CIRCUIT BREAKER AND DISTRIBUTION PANEL
GFCI
4 GFCI

20A 15A 15A 15A 15A 15A 20A 20A 20A 15A

20A 20A 20A 20A 30A 50A 20A 15A 20A 50A GFCI

GARAGE OUTLETS — Attic Fan, Outdoor Lights
FURNACE — Space Heater 1600W Soldering Iron 150W
SHOP EQUIPMENT — Drill 500W Lathe 300W Saw 500W
UTILITY ROOM — Iron 1100W Washer 700W
UTILITY ROOM — Clothes Dryer 5200W
OVEN - RANGE — 8200 to 16,000W
DISHWASHER — to 1800W
SINK DISPOSAL UNIT — to 900W
WATER HEATER — 2500W
CENTRAL AIR CONDITIONER — 10,000W

FIGURE 22B. Circuits in the completed residential wiring system are sized to meet the requirements of family living and leisure.

NOTES

II. Installing Wiring

Wiring procedures discussed in this section are those used to install **nonmetallic sheathed cable** (Figure 23). Local codes in some areas may require the use of other kinds of material such as conduit (metal or plastic) (Figure 24). Flexible armored cable may also be used (Figure 25).

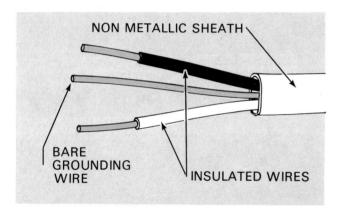

FIGURE 23. Nonmetallic sheathed cable is most often used for house wiring.

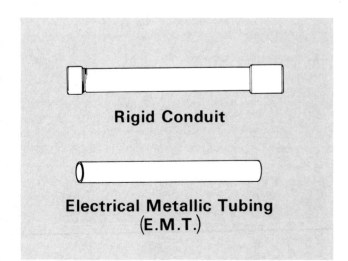

FIGURE 24. Some local codes require the use of conduit.

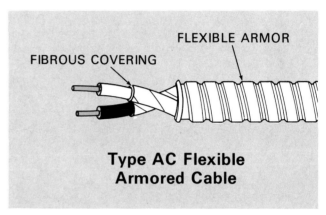

Type AC Flexible Armored Cable

FIGURE 25. Flexible armored cable is required by some local codes.

Nonmetallic sheathed cable is most often used for house wiring (Figure 23). If you learn the correct procedures for installing nonmetallic sheathed cable, you will be able to install other kinds of wiring with a minimum of additional instruction.

Upon successful completion of this section, you will be able to **select** the **correct type** and **size** of **cable** and **other material required, locate** and **mark** the **location** for the **service entrance panel** and **lo-**

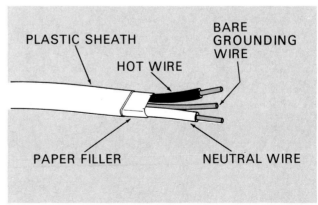

FIGURE 26. Nonmetallic sheathed cable is made up of plastic sheathing, paper filler, insulated wires and a bare grounding wire.

cate and **mark routes** for **all types** of **circuits.** You will be able to **install outlet boxes, switch boxes** and **junction boxes, provide access** for cable, and **pull** and **install cable** to **all types** of **outlets.**

Installing wiring is discussed under the following headings:
A. Observing Safety Practices.
B. Selecting Tools and Equipment.
C. Locating Circuits.
D. Installing Device Boxes and Outlet Boxes.
E. Providing Access for Nonmetallic Sheathed Cable.
F. Pulling Cable for Circuits.
G. Anchoring Cable.

A. Observing Safety Practices

In addition to following safety requirements of the Code, there are some common sense safety rules that should be followed at all times. The electrician should develop safe working habits that will protect against accidents common to many workers. Some of these practices are:

1. *Wear appropriate clothing, shoes and headwear for the job to be done (Figure 27).*

   A **hard hat** and **steel-toed shoes** will help protect the worker from falling objects. **Rubber heels** and **soles** without nails help to insulate against shock. **Clothing** should **fit well** and be free of flapping pockets that might be caught on projections. Do not wear metal rings and watchbands that might get caught on something. Wear **safety glasses** when doing work where flying particles or fluids might cause problems.

FIGURE 27. Wear appropriate clothing, shoes and hat for personal safety.

2. *Use tools that protect you.*

   Use only **UL listed power tools. Double insulated power tools** offer valuable protection against electric shock (Figure 28). Hand tools such as screwdrivers and pliers should have **insulated handles** for further protection against shock.

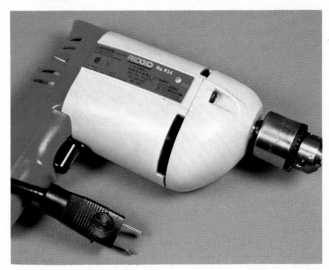

FIGURE 28. Use only UL listed tools, preferably double insulated.

3. *Keep tools in good condition.*

   Cutting tools should be kept sharp. Any damaged or broken tool should be repaired or replaced. Keep screwdriver blades correctly ground and use the correct size. Be sure all cords and plugs on non-double insulated power tools are grounding type and in good condition. Keep them in good condition by removing plugs correctly and by protecting cords. Grasp the plug and pull directly on the plug to remove it. Never jerk on the cord. If extension cords must cross a traffic area, protect them with planks (Figure 29) or other means.

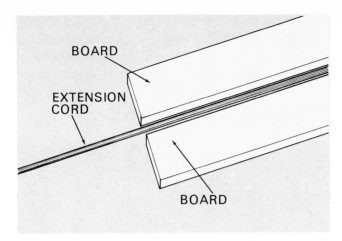

FIGURE 29. Protect extension cords if used in a traffic area.

FIGURE 31. Keep your work area clear to prevent falls.

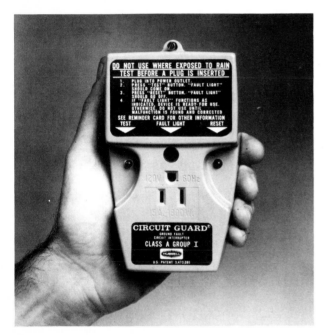

FIGURE 30. Use portable Ground Fault Circuit Interrupters if the power source is not protected permanently at the service entrance.

DO NOT USE TOP TWO STEPS ON LADDER

FIGURE 32. Use ladders safely.

4. *Use Ground Fault Circuit Interrupter for power tools.*

   Portable Ground Fault Circuit Interrupters (GFCI) are available that may be moved to any outlet for protection (Figure 30). If the power source is not protected by GFCI, the portable type is required for construction sites.

5. *Keep the work area clear (Figure 31).*

   Developing good housekeeping habits can prevent injuries caused by tripping and falling.

6. *Use ladders safely (Figure 32).*

   Use only wooden or fiberglass ladders for electrical work.

   Check ladders regularly to see that they are in good condition. Repair loose or broken rungs immediately. Use ladders with rubber or non-slip feet. Have someone hold the ladder at the base if necessary. Be sure no tools are left on step ladders and do not use the top two steps to stand. Be sure all braces are locked in place.

7. *When working in a damp location, take precautions against shock (Figure 33).*

   Don't stand on wet ground or a damp floor when using power tools without protecting yourself. Stand on a rubber mat or other non-conducting material for protection.

8. *Never work on a "hot" electrical circuit (Figure 34).*

   Disconnect power to the branch circuit before you work on wiring or equipment. To do this, turn the main service-entrance switch off. Then lock the switch cover in place to prevent someone from turning on the switch while you are working on the electrical system. If for some reason this cannot be done, disconnect power to the branch circuit where you will be working. Remove the branch-circuit fuse(s) or turn the circuit breaker to the "off" position. Then test the circuit to be sure it is not live by plugging in a lamp or circuit tester (Figure 35).

FIGURE 34. Disconnect the circuit to be worked on and lock the cover whenever possible.

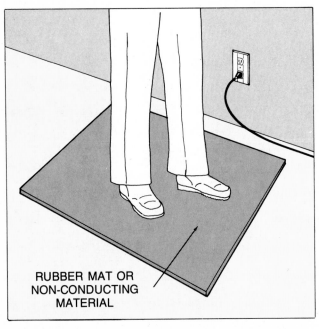

FIGURE 33. A rubber mat or other non-conducting material helps protect you on wet ground or damp basement when using power tools.

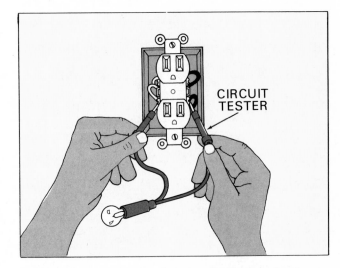

FIGURE 35. Test the circuit to see if it is energized.

B. Selecting Tools and Equipment

Basic tools needed by the electrician are (Figure 36):

—**Screwdrivers.**

Flat blade Phillips and Robertson in several sizes.

—**Pliers.**

Pliers of various types probably rank first among tools frequently used by electricians. **Linemans** pliers are available in several sizes with 18 cm and 20 cm (7″ and 8″) sizes being most popular. The flat jaws will bend, grip, pull and twist light or heavy cable and wires. The side cutting jaws will cut cable up to large sizes and the shoulders will crush insulation to make cutting easier. **Locking** pliers are used to grip and hold cable, tighten connectors and pull cable. **Diagonal cutters** have exposed tapered jaws that are convenient for cutting cable ends in boxes and other places where close fitting connections are made. **Long nose pliers** (often called needle nose) are useful for bending loops in wire to fit under screws. **Round nose pliers** make a desirable rounded loop in wire to fit under screws.

Water pump (channel locking) pliers are used to tighten locknuts and cable connectors.

All pliers should have insulated handles. Slip-on insulated covers are available for most sizes.

—**Wire stripper.**

This tool quickly removes insulation without damage to the wire. Some types include wire cutting jaws.

—**Wrenches.**

Adjustable jaw should have several sizes—20 cm and 25 cm (8″ and 10″) are preferred by many. **Pipe wrench** is used to grip and turn conduit.

—**Drilling tools.**

Portable electric drills and **carpenters brace and bits** are used to drill holes. Several sizes and types of bits may be used. Hole-saw bits may be used to make larger holes. Various types of auger bits are used for drilling holes in framing. Carbide-tipped bits are required to drill into masonry. Also, star masonry hand bits are often used.

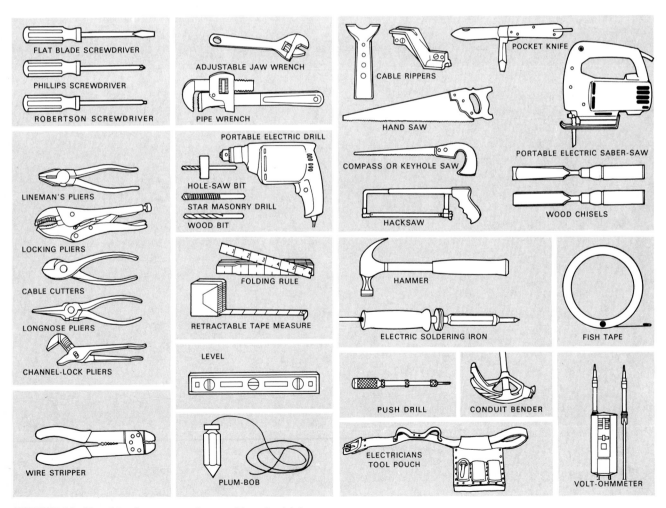

FIGURE 36. Hand tools commonly used by electricians.

—**Measuring tools.**

Folding extension rules made of wood, metal or plastic are usually six or eight feet in length. The six-inch extension slides out to measure box set-out. Metal tapes with spring return are 1.8 to 6.1m (6 to 20 feet) in length. Fifteen-meter or 30-meter (50-foot or 100-foot) **tapes** may be metal or fabric and usually have a crank return. The combination square is useful for simple squaring.

—**Level.**

These come in many sizes. They are useful for horizontal leveling or for vertical plumb position.

—**Plumb bob.**

Line with weight attached to find true perpendicular line.

—**Cutting tools.**

The cable ripper will quickly slit the sheath (outer jacket) on cable. A heavy duty **pocket knife** may also be used to slit sheath and may be used to taper insulation on wire ends. The standard **carpenters crosscut saw** is for sawing wood. The **compass saw** or keyhole saw is a small tapered saw for rough cutting holes in flooring, paneling and plasterboard. The **hacksaw** is used for metal cutting, including large cable or conductors. **Wood chisels**

are used to cut or gouge wood. The **electric saber saw** is very versatile. It may be used for cutting wood, metal and plasterboard.

— **Hammer** is used for nailing boxes and other devices in place and for driving staples.

—**Soldering iron.**

Soldered connections are not now required in most localities but may be called for under some conditions. Soldered connections are not allowed for certain connections.

—**Tool pouch.**

Use a belt-mounted leather pouch with space to carry "most-used" tools.

—The **fish tape** is used to pull or push wire through conduit.

—The **Volt-Ohm meter** is used to test circuits for voltage and continuity.

—**Crimping tool.**

Available in several sizes, they are used for crimping connectors on wires.

C. Locating Circuits

On most wiring jobs, you will have a plan showing the location of electrical outlets in the house. You must decide how to route the branch circuits from the service entrance panel to each outlet shown on the plan (Figure 37). All circuits discussed in this section are branch circuits.

Locating circuits is discussed under the following headings:

1. Marking Location for Service Entrance Panel.

2. Locating and Marking Routes for Small Appliance Circuits.

3. Locating and Marking Routes for General Purpose Circuits.

4. Locating and Marking Routes for Individual Circuits.

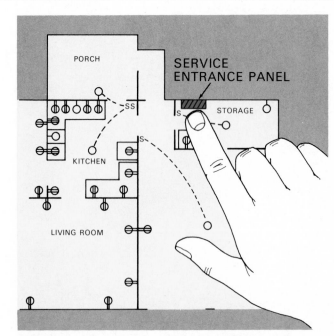

FIGURE 37. Circuits must be routed from the SEP to outlets shown on the plan.

1. MARKING LOCATION FOR SERVICE ENTRANCE PANEL

The procedures described in this section are part of the "roughing-in" process in building a house. Proceed as follows:

1. *Locate position of SEP (Figure 38).*

   The SEP should be located as near the service entrance as possible. Check with your power supplier to determine where the electric service will enter the house. It is also desirable to locate the SEP near the appliances that use the most power. The reason is that these appliances require large conductors which are expensive. They include kitchen appliances, electric water heater, clothes dryer and electric furnace.

   The SEP must be easily accessible at all times. Keep in mind the future use of the area when locating the panel. The Code prohibits locating the SEP in a clothes closet.

2. *Mark the location selected on a stud or wall using pencil or crayon (Figure 39).*

   Mark the location of the SEP cabinet for either surface mount or flush mount as required. The bottom edge of the cabinet should be about 102 – 122 cm (40 – 48 inches) above floor level or as required.

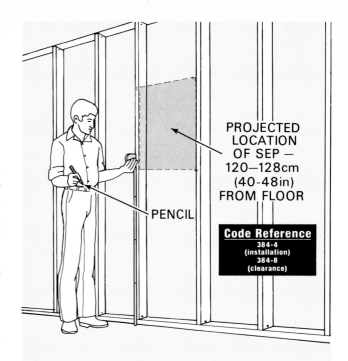

PROJECTED LOCATION OF SEP — 120–128cm (40-48in) FROM FLOOR

PENCIL

Code Reference
384-4 (installation)
384-8 (clearance)

FIGURE 39. Using crayon or pencil, mark the location of the SEP.

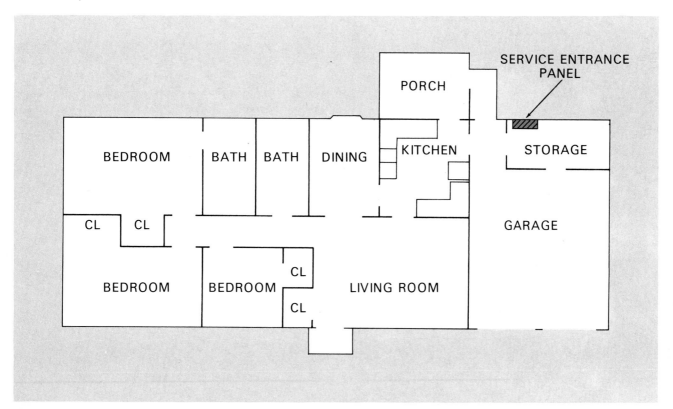

FIGURE 38. Locate the SEP as near as possible to the point where the service entrance conductors enter the house.

2. LOCATING AND MARKING ROUTES FOR SMALL APPLIANCE CIRCUITS

When you first begin locating and marking circuits, it may be helpful to indicate the location of each one on your floor plan similar to those used in this manual. As you gain experience you will be able to mark and follow circuit routes on the framing members.

There are several different ways to route (run) circuits in most wiring jobs and each one may be satisfactory. The route for **small appliance** circuits should be fairly easy to locate if Code requirements are followed.

Locating and marking routes for small appliance branch circuits are discussed under the following headings:

    a. Locating and Marking Receptacles.
    b. Locating and Routing Small Appliance Circuits.

a. Locating and Marking Receptacles

Your house plan shows where each receptacle outlet is to be located. Mark each of the outlets on the stud where the box is to be mounted (Figure 40). Each

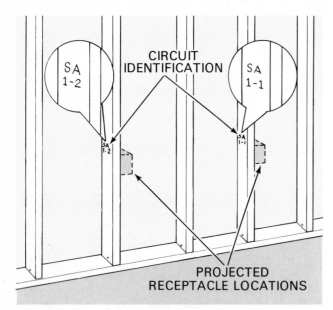

FIGURE 40. Mark the location of each receptacle using your house plan as a guide and identify the circuits. Mark the stud small appliance 1 (SA-1-1) for the first outlet, (SA-1-2) for the second outlet etc.

receptacle on the kitchen counter should be located so the bottom of the box will be approximately 20 cm (8 inches) above the counter top. All other receptacles should be 30 to 36 cm (12 to 14 inches) from the floor. Some people use the length of the hammer handle as a quick measure of correct height.

b. Locating and Routing Small Appliance Circuits

The following example is based on the plan shown in Figure 41. Proceed as follows:

1. *Determine number of small appliance circuits needed.*

   On the plan in Figure 41, only the kitchen and dining room have small appliance outlets, so two circuits should be enough. The small appliance circuit shown in red is divided between the kitchen outlets and the dining room outlets. Small appliance circuit No. 2 contains only kitchen receptacles. The laundry circuit is also shown in black in Figure 41.

2. *Locate and mark small appliance circuit No. 1.*

   For example, in Figure 41, circuit No. 1 is shown in red. A circuit or part of a circuit between outlets is often referred to as the **circuit run.** In some instances, it will be called a cable run or just a run.

   (1) *Mark circuit run beginning at kitchen receptacle outlet near the door.*

   Using a crayon, identify circuit No. 1 on the studs near the locations of the receptacles (Figure 40). For example, starting with receptacle No. 1 (SA1–1), receptacle No. 2 (SA1–2) and continue until all outlets on the circuit have been identified.

   Route the cable one of two ways to reach the outlets. Run either overhead through the attic or under the floor and up.

   (2) *Mark the circuit run to outlet No. 2 above the countertop.*

   (3) *Mark circuit run to outlet No. 3.*

   From the second outlet, run the cable through the studs above the countertop to the third outlet (Figure 42).

   (4) *Mark circuit run to first and second dining room outlets.*

   Run the cable through the stud to the first dining room outlet and overhead to the second dining room outlet.

   (5) *Mark circuit run to dining room outlet No. 3.*

   To reach the third outlet across the room, again run either overhead through the attic or

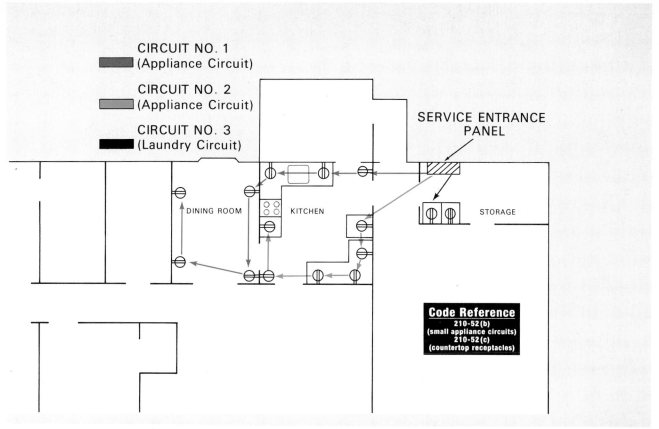

CIRCUIT NO. 1
(Appliance Circuit)

CIRCUIT NO. 2
(Appliance Circuit)

CIRCUIT NO. 3
(Laundry Circuit)

SERVICE ENTRANCE
PANEL

DINING ROOM   KITCHEN

STORAGE

Code Reference
210-52(b)
(small appliance circuits)
210-52(c)
(countertop receptacles)

FIGURE 41. Small appliance circuits Nos. 1 and 2 and the laundry circuit. Counter-top receptacles must be installed so that no point is more than 24 inches away from a receptacle.

under the floor. Figure 43 shows the cable run from the attic through the plate down to the third dining room outlet, and through the studs to dining room outlet No. 4. This completes small appliance circuit No. 1.

3. *Locate and mark small appliance circuit No. 2.*

Small appliance circuit No. 2 is shown in blue (Figure 41). Identify the circuit on the stud as described for circuit No. 1.

NOTE: Counter-top receptacles must be positioned so that no point along the wall line of the counter is more than 24 inches from the nearest receptacle. Island or peninsular counter tops 12 inches wide or wider must have at least one receptacle for each four feet of countertop.

(1) *Run small appliance circuit No. 2 from the service entrance panel overhead to the refrigerator outlet.* (Some prefer to place the refrigerator on a separate circuit.)

(2) *Run through studs to the other three outlets above the counter top.*

(3) *From the last counter-top outlet, run the cable overhead and down to the wall outlet beside the door.*

To reach the outlet above the counter top across the second doorway, run the cable overhead and down to the outlet to complete circuit No. 2.

4. *Locate and mark the laundry circuit.*

Run the laundry circuit overhead to the outlet for the clothes washer (Figure 41). The clothes dryer is often connected to a 120/240-volt circuit so it must be wired separately. It is discussed under Individual Circuits on page 34.

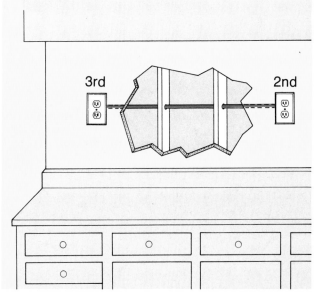

3rd   2nd

FIGURE 42. Small appliance circuit No. 1 runs through studs above the kitchen counter to the third outlet.

27

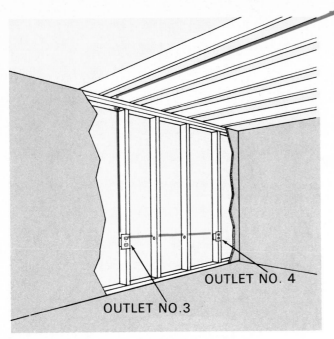

FIGURE 43. Cable routed overhead from the attic to dining room outlets three and four.

3. LOCATING AND MARKING ROUTES FOR GENERAL PURPOSE CIRCUITS

General purpose branch circuits include lighting outlets, and most convenience outlets, such as those used for TV, radio, stereo, vacuum cleaners, and hair dryers.

Figure 44 shows a suggested layout for the six general purpose circuits. Several other combinations are possible that would work as well. Each circuit begins at the outlet most convenient to the SEP as indicated by the arrow.

No. 12 wire is often required as the minimum wire size for general purpose circuits by local codes. However, the Code permits the use of No. 14 wire for general purpose circuits.

In this section, you will lay out and mark a practical and economical route for each of the six general purpose circuits in the wiring plan (Figure 44). Keep in mind that the target number is 10 outlets or less per circuit for No. 14 wire and 13 outlets or less for No. 12 wire. Mark

General Purpose Circuits

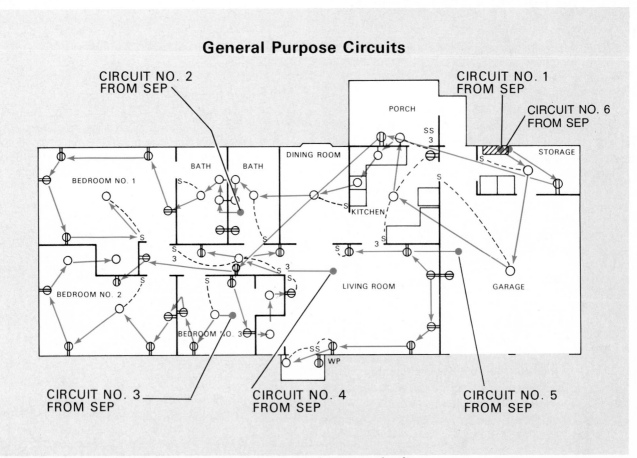

FIGURE 44. The standard house plan contains five general purpose circuits.

28

the switch and receptacle locations as each circuit is completed. Switch height should be 122 cm (48 inches) from floor to center of box and on the latch side of the door. Some outlets may have more than one switch.

NOTE: Code Section 210-70A requires that where more than six steps separate floor levels in a dwelling, a wall switch must be installed at each floor level for interior stairway lighting.

Locating and marking routes for general purpose circuits are discussed as follows:
a. Marking Location for Overhead Lighting Outlets.
b. Locating Circuit No. 1.
c. Locating Circuit No. 2.
d. Locating Circuit No. 3.
e. Locating Circuit No. 4.
f. Locating Circuit No. 5.
g. Locating Circuit No. 6.

a. Marking Location for Overhead Lighting Outlets

Overhead lighting outlets are usually located in the center of the room. To mark the location for the overhead lighting outlet, proceed as follows:

1. *Use a chalk line to establish a diagonal line from one corner to another on floor.*

2. *Repeat for the other two corners.*

3. *Mark the spot where the lines cross (Figure 45).*

4. *Locate the ceiling outlet above this point by means of a plumb bob, or extend a ruler upward to mark the spot on a ceiling joist.*

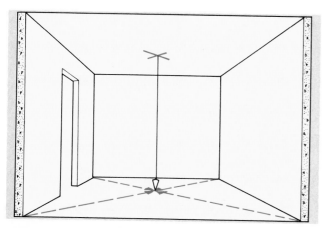

FIGURE 45. Draw diagonal lines from each corner to locate lighting outlet in center of room.

b. Locating Circuit No. 1

The lighting outlets for the storage room, garage, bathroom, kitchen and dining room are part of general purpose circuit No. 1. Locate and mark each outlet and switch on the circuit.

Some of the circuit runs could be under the floor if you have a crawl-space or basement. Other parts of the circuit may be overhead or through the studs. Proceed as follows (Figure 46):

1. *From the SEP, run the circuit to the ceiling light outlet in the storage room.*

   Identify the circuit by marking the outlets and receptacle locations in sequence. For example, (GP1-1, GP1-2, GP1-3, etc.)

2. *Extend to the garage ceiling light.*

3. *From the garage, run the circuit to the ceiling light in the kitchen.*

4. *From the kitchen ceiling light, run the circuit to the wall light on the porch.*

5. *Return to the other two kitchen lights.*
   (1) *Run the circuit first to the light above the sink.*
   (2) *Extend to the light above the range.*

6. *Continue the circuit to the dining room ceiling light.*

7. *From the dining room, run to the ceiling light in bathroom No. 1.*

8. *From the bathroom ceiling light, extend the run to the two wall lights, one on each side of the bathroom mirror.*

   The circuit ends at the light on the left. This completes a circuit containing 10 outlets.

9. *Mark location of switches.*

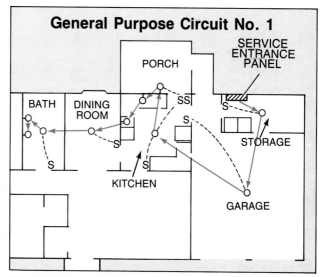

FIGURE 46. Beginning at the SEP, general purpose circuit No. 1 extends from a lighting outlet in the storage room to others in garage, kitchen, porch, dining room and bathroom No. 1.

c. Locating Circuit No. 2

Circuit No. 2 (Figure 47) shows light outlets and convenience outlets mixed on the same circuit. Check your local code to see if both types of outlets are permitted on the same branch circuit.

In circuit No. 2, some of the circuit runs could be under the floor, overhead or through the studs (Figure 48).

The arrow extending from the wall lighting outlet in bathroom No. 2 indicates the starting point for circuit No. 2 (Figure 47). To show the entire circuit would require a line drawn from the SEP to the bathroom outlet.

Proceed as follows to locate circuit No. 2:

1. *Beginning at the wall light in the bathroom, extend the circuit to the wall light on the opposite side of the mirror.*

   Identify circuit by marking the location of each outlet and receptacle as follows: GP2-1, GP2-2, GP2-3, etc.

2. *Run next to the combination ceiling light and fan (Figure 47).*

   Note: A combination ceiling light, fan and heater may require a separate circuit, depending on heater wattage.

3. *From the bathroom ceiling light, run to the nearest bedroom receptacle.*

4. *The remaining receptacles are then connected in order.*

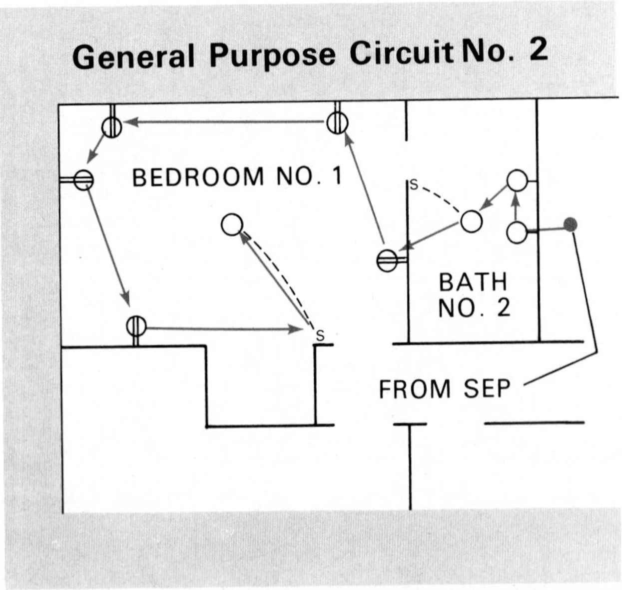

(1) *Make the run between the two outlets on the same wall through the studs (Figure 48).*

(2) *Run the other two through the studs or overhead (Figure 48).*

5. *From the last receptacle, run to the switch and from there to the ceiling light.*

6. *Mark location of switches.*

FIGURE 48. Cable run through holes in studs.

d. Locating Circuit No. 3

Circuit No. 3 also includes both ceiling lights and receptacles (Figure 49). Note that it includes all out-

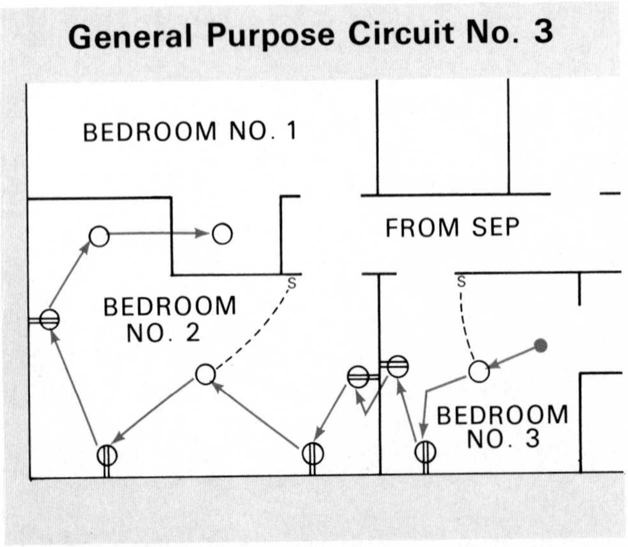

General Purpose Circuit No. 2

BEDROOM NO. 1

BATH NO. 2

FROM SEP

FIGURE 47. Circuit No. 2 includes lighting outlets and receptacles in bathroom No. 2 and bedroom No. 1. The starting point of circuit No. 2 is indicated by the arrow at the bathroom wall lighting outlet.

General Purpose Circuit No. 3

BEDROOM NO. 1

FROM SEP

BEDROOM NO. 2

BEDROOM NO. 3

FIGURE 49. Circuit No. 3 includes lighting outlets and receptacles in closets and in bedroom No. 2 and part of bedroom No. 3.

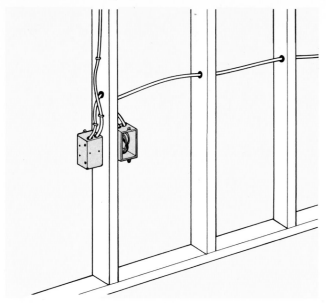

FIGURE 50. Back to back receptacles on opposite sides of the stud serving different rooms.

lets in one bedroom and part of another. Proceed as follows:

1. *From the SEP, start with the ceiling light outlet indicated by the arrow.*

    Identify circuit No. 3 at each outlet or receptacle location.

2. *Run next to a bedroom receptacle.*

3. *From there extend to a second outlet that is back-to-back with another in the adjoining bedroom (Figure 50).*

    This is a common arrangement that saves cable.

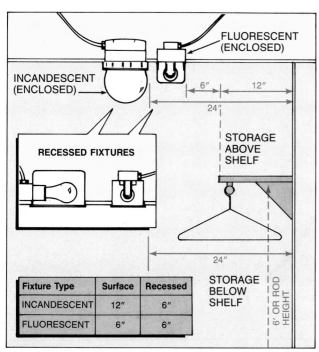

FIGURE 50A. There are guidelines regulating the placement of closet lighting fixtures in relation to storage areas.

4. *Run to a second receptacle in the bedroom (Figure 49).*

5. *Run to the ceiling light outlet.*

6. *Extend to the other two receptacles.*

7. *Add the ceiling lights in the two closets to complete the circuit containing 10 outlets.*

    Incandescent lights mounted in closets must have 12 inches clearance between the fixture and the storage area (Figure 50A). Recessed flourescent and incandescent fixtures may be mounted on the wall or ceiling with 6 inches of clearance between the fixture and the storage area. These lamps must be completely enclosed. Pull chain-type lampholders are prohibited in clothes closets.

    Section 410-8 of the Code provides the guidelines concerning types of fixtures allowed and their placement in clothes closets.

8. *Mark location of switches.*

---

e. Locating Circuit No. 4

Circuit No. 4 is the most compact of all. It covers the hall light and outlets, the remainder of bedroom No. 3 and one living room receptacle (Figure 51). Proceed as follows:

1. *Begin with the hall ceiling light switch at the arrow.*

    Identify circuit No. 4 at each outlet and receptacle location.

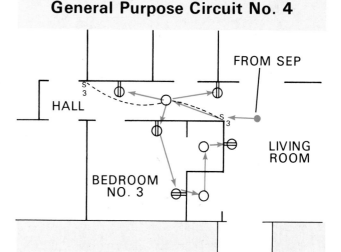

FIGURE 51. Circuit No. 4 serves the hall light and outlets, part of the bedroom and one living room receptacle.

2. *Run to the hall ceiling outlet.*

    Here you find a situation not previously covered. From the hall ceiling outlet the circuit runs in three directions. This is a common arrangement that you will use often.

3. *Extend one run to the first receptacle in the hall. Extend another to the second hall receptacle.*

4. *Run next to the nearest receptacle in bedroom No. 3.*

5. *From the receptacle, run to the second bedroom receptacle.*

6. *Extend to the two closet ceiling lights.*

7. *Run to the receptacle on the living room wall. This complete circuit contains 8 outlets.*

8. *Mark location of switches.*

---

f. Locating Circuit No. 5

Circuit No. 5 is made up of the receptacles in the living room plus one wall light outlet outside the front door and one in the garage (Figure 52). Proceed as follows:

1. *Begin the circuit at the receptacle indicated by the arrow.*

    Identify circuit No. 5 at each outlet and receptacle location.

    The circuit runs two ways from the starting point.

2. *Run next to the two outlets near the kitchen and dining room doors.*

3. *On the other side, run to the remaining receptacles in living room and garage.*

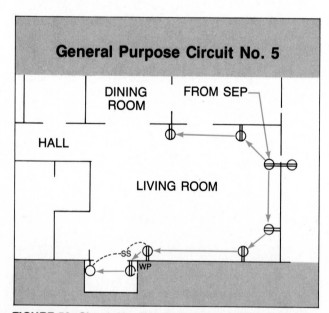

General Purpose Circuit No. 5

FIGURE 52. Circuit No. 5 includes the receptacles in the living room plus the outside light and receptacles near the front door and in the garage.

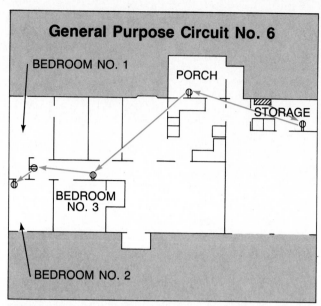

General Purpose Circuit No. 6

FIGURE 52A. Circuit No. 6 includes a receptacle in the storage room, one on the porch, two in the hall, and one in bedroom No. 2. It also provides capacity for additional outlets as needed.

4. *Extend to the outside receptacle and wall light.*

5. *Mark location of switches.*

This completes a layout of 5 of the general purpose circuits. As you have seen, none of the circuits has more than 10 outlets. Circuit Nos. 1 and 3 have exactly 10. The others contain 9, 8 and 9, leaving little room to add outlets if needed at some future date.

---

g. Locating Circuit No. 6

Circuit No. 6 contains five receptacles — one located in the storage room, one on the porch, two in the hall and one in bedroom No. 2 (Figure 52A). Proceed as follows:

1. *From the SEP, run to receptacle in the storage room opposite the water heater.*

2. *Extend to the porch receptacle.*

3. *From the porch extend to the two hall receptacles, one on the wall of bedroom No. 3 and the other at the end of the hall.*

4. *Extend to the receptacle in bedroom No. 2 on the closet wall.*

This centrally located circuit provides a convenient way to extend wiring to any part of the house by installing required junction boxes as explained on page 90.

4. LOCATING AND MARKING ROUTES FOR INDIVIDUAL CIRCUITS AND FOR GROUND FAULT CIRCUIT INTERRUPTERS

Locating and Marking Routes for Individual Circuits and for Ground Fault Circuit Interrupters are discussed under the following headings:

    a. Locating and Marking Routes for Individual Circuits.

    b. Locating and Marking Routes for Ground Fault Circuit Interrupters.

a. Locating and Marking Routes for Individual Circuits

Individual branch circuits are installed for *heavy duty* appliances and equipment. According to the Code, each must have its own circuit (Figure 53). This means that a separate cable is run from the SEP to the outlet for the heavy duty item. Some are connected direct without use of plug or receptacle. Another difference in some individual circuits is the higher voltage. All small appliance and general purpose circuits are wired for 120 volts. Several items on individual circuits require 240 volts or a combination of 120/240 volts. The 240-volt circuits or 120/240-volt circuits may call for larger wire sizes and higher capacity fuses or circuit breakers. Some common individual circuits are indicated by arrows.

Individual circuits for the following items are usually wired at 120 volts with **No. 12 wire** and protected by a **20-ampere** fuse or circuit breaker.

—**Garbage disposer.***
—**Dishwasher.***
—**Trash compactor.**
—**Motor on fuel-fired furnace.**
—**Room air conditioner.**
—**Bathroom heaters.**

*Many local codes allow the garbage disposer and dishwasher to be wired on the same circuit.

The items listed above are usually wired on individual circuits at 240 volts or 120/240 volts. Wire sizes and fuse or circuit breaker sizes as shown in the chart are common examples (Table III).

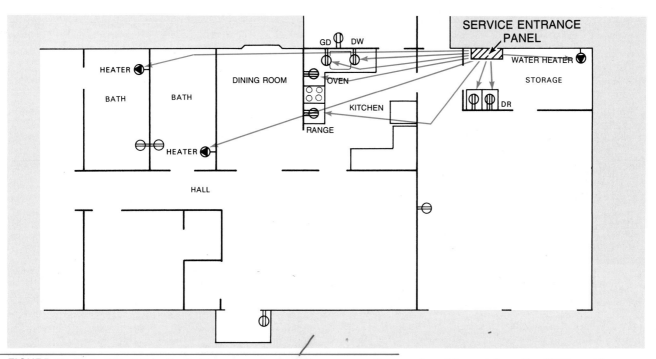

FIGURE 53. Each circuit is indicated by a receptacle with an arrow. A separate cable runs from the SEP to each receptacle so marked. Receptacles not marked with an arrow are protected by a GFCI, which may be installed at each outlet or in the SEP.

TABLE III. EXAMPLES OF VOLTAGES, WIRE SIZES AND OVERCURRENT PROTECTION FOR INDIVIDUAL CIRCUITS

Usual Voltage	Circuit	Copper Wire Size (AWG)	Fuse or Circuit Breaker (Amps)
240	Air conditioner (large)	12	20
240	Water heater	12/10	20/30
120/240	Separate oven	10	30
120/240	Counter top cooking unit	10	30
120/240	Clothes dryer	10	30
120/240	Self contained range	6	50
240	Central air conditioning or heat pump	8 or 6	40–60
240	Water pump	12	20
240	Central electric furnace	6	50

Food freezers may also be on a separate circuit to eliminate the possibility of a power loss due to other items on the circuit. However, it is a good idea to include **another outlet** on the freezer circuit to indicate power loss, such as a clock or a light.

When marking the locations for each of the special purpose outlets, you should identify the 120/240-volt or 240-volt circuits. A different size and type of outlet is required for each.

The individual circuits in Figure 53 are as follows:

1. *Beginning in the storage room, locate the water heater on an individual 240-volt circuit.*
   Identify individual circuits by marking the outlet location, such as (WH).
2. *Locate the dryer on an individual 120/240-volt circuit.*
   The **washer**, previously mentioned, is on a 120-volt circuit.
3. *In the kitchen, place the separate oven and countertop cooking unit on 120/240-volt circuits.*

FIGURE 54. Type of GFCI installed at outlet.

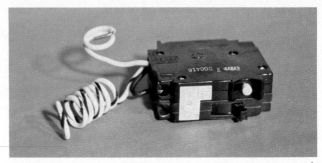

FIGURE 55. Type of GFCI used in service entrance panel.

4. *Locate both the dishwasher and garbage disposal on 120-volt circuits, (or combine if local code permits).*
5. *Locate the bathroom heaters normally on 120 volts. They may be connected to 240 volts if so designed.*

b. Locating and Marking Routes for Ground Fault Circuit Interrupters

In addition to the individual circuits already mentioned, many people prefer to run individual circuits to outside receptacles. An example is one near the front door and one in the garage. **Outdoor receptacles located within 6'6" of finished grade and all kitchen receptacles within 6' of the kitchen sink shall have GFCI protection**, according to Code Section 210-8 (a) (3). Code Section 210-8 (a) (4) requires that all outlets in an unfinished basement and crawl spaces (except for specific exceptions) be GFCI-protected. All outlets in a garage must be GFCI-protected (except for specific exceptions), according the Code Section 210-8 (a) (2). Receptacles for plug-connected garage door openers and receptacles intended for food freezers, sump pumps, and laundry equipment in a garage are not required to have GFCI protection. The 1993 Code requires a receptacle at both the front and rear entrance. These receptacles must also be GFCI-protected, according to Code Section 210-8 (3). The purpose of the GFCI is to provide protection for people who may be using a faulty tool or equipment or may accidentally misuse a tool that could cause a dangerous shock. A shock received when standing on wet ground can be severe enough to cause death. The GFCI device may be installed at each outlet (Figure 54), or in the SEP (Figure 55).

Each convenience outlet in the bathroom must also be protected by a GFCI. Bathrooms are defined as an area having a basin and one or more of the following: a toilet, a tub or a shower. In the house in Figure 53, there is one in each bath. One circuit to the two outlets should be satisfactory. Both may be protected from a single GFCI device installed in the service entry panel. Procedures for GFCI installation are discussed in "**Section IV. Installing Service Entrance Equipment.**"

D. Installing Device Boxes and Outlet Boxes

Installing device boxes and outlet boxes is often among the first tasks given the beginning electrician. **A device is a unit of the electrical system which is intended to carry, but not use, electricity.** Examples of devices are switches and receptacles (Figure 56).

FIGURE 56. Switches and receptacles are examples of the devices installed in device boxes.

Some electricians prefer to drill the holes for the cable before mounting the boxes. This is a matter of personal preference and can be done either way. **All connections and splices must be made in a box.** Your house wiring plan shows you where each box must be mounted.

Installing device boxes and outlet boxes is discussed under the following headings:
1. Types and Sizes of Device Boxes and Outlet Boxes.
2. Number of Conductors Allowed in Boxes.
3. Installing Switch Boxes and Handy Boxes.
4. Installing Outlet Boxes.

1. TYPES AND SIZES OF DEVICE BOXES AND OUTLET BOXES

Most boxes used in house wiring are made of plastic (Figure 57). Another type of box made of galvanized sheet steel is used for special applications in residential wiring (Figure 58). Plastic boxes (nonmetallic) may only be used with nonmetallic sheathed cable or nonmetallic (plastic conduit) (Figure 58).

Nonmetallic boxes are installed almost exclusively in a residential installation. They are available in a variety of sizes and shapes for most all applications. The cost of these boxes is relatively cheap compared to conventional metallic boxes.

FIGURE 57. Nonmetallic boxes made of durable plastic.

Another factor of major importance is that a non-metallic box requires no cable clamps inside the box. This greatly increases the available space inside the box. It should be noted that *NEC*® Table 370-16 (a), which has information on the maximum number of conductors in a box, is applicable to metal boxes only.

Nonmetallic boxes are available in both 80 degree and 90 degree C. temperature ratings. These temperature ratings will apply when special conditions exist where an ampacity rating is based on the cables' insulation temperature as per Code table 310-16. In a residential installation, a plastic or PVC box is available for virtually all applications.

FIGURE 58. Metal boxes made of galvanized sheet steel.

One application where a nonmetallic box cannot be used is for an overhead paddle fan, according to *NEC*® 410-16(a). The constant torqueing, twisting, and vibrations of a fan require that a special box be used. These boxes must be UL-approved for this application. There are further restrictions based on the weight of the device that is mounted. See *NEC*® 370-27 (c) for additional information.

Boxes of three basic shapes and many sizes are available for house wiring.

Types and sizes of device boxes and outlet boxes are discussed as follows:
   a. Types and Sizes of Device Boxes.
   b. Types and Sizes of Outlet Boxes.

a. Types and Sizes of Device Boxes

Types and sizes of device boxes are discussed as follows:

(1) Types of Device Boxes. The box in Figure 59 is a **device** box. It is usually called a **switch box,** but it is more than that. It is used to install switches and for several other purposes. For example, you will use more of them to install duplex receptacles than ·switches. They are referred to as switch boxes in the following discussion.

Switch boxes are available with square or beveled corners and with mounting brackets of various types (Figure 60). Other types are those with brackets on top and bottom with 16d nails attached (Figure 61) and those with holes through the sides for nails (Figure 62).

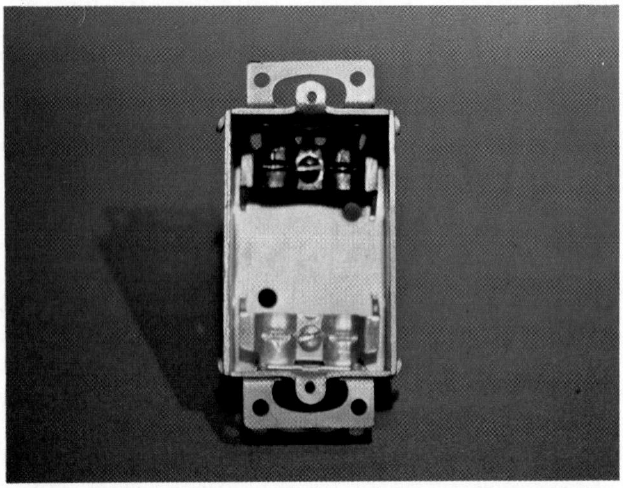

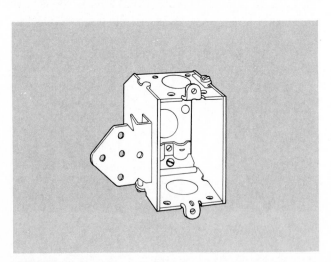

FIGURE 59. A common type of device box, usually called a switch box.

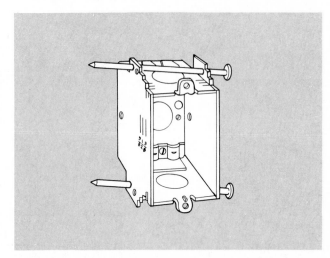

FIGURE 61. A switch box with square corners and nails through brackets at bottom and top for mounting.

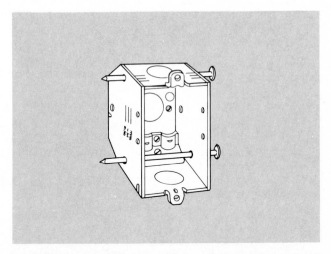

FIGURE 60. A device (switch) box with beveled corners and mounting bracket.

FIGURE 62. A switch box with beveled corners and nail holes through the sides of the box for mounting.

Detailed features of the switch box are seen in Figure 63. **Knockouts** are the coin-sized plugs shown on the sides, bottom and rear of the box. A sharp blow knocks out the plug which has been cut out except for a narrow strip (Figure 64). Removal of the plug makes an opening where needed to bring in cable or conduit. **Knockout openings that are not used must be plugged with knockout seals** (Figure 64 inset).

Pryouts (rear, top and bottom) are smaller knockouts that are also present on some boxes. They are for nonmetallic cable installation and are always in pairs. Remove those needed by inserting a screwdriver in the slot, then pry up and twist (Figure 65).

Mounting ears make mounting and adjusting easier. They are especially useful when installing a switch box in an existing (older) home (Figure 66).

FIGURE 63. Knockouts are the coin-size circles on the sides, bottom and rear of the box.

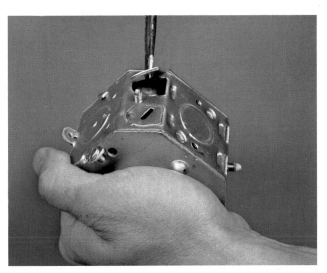

FIGURE 65. Pryouts are smaller openings present on some boxes that are removed by prying and twisting with a screwdriver.

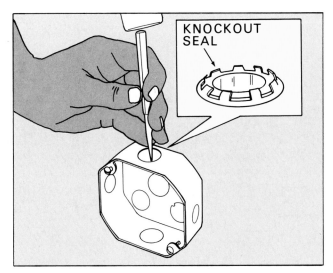

FIGURE 64. A sharp blow knocks out the plug to make an opening. The knockout seal is used to plug unused openings.

FIGURE 66. Mounting ears are used to mount and adjust the box.

The Code now requires that a way be provided for connecting a grounding wire in each metal box. In most boxes, this will be a tapped (threaded) hole that will accept a screw for the grounding terminal (Figure 67).

The pre-installed **pigtail grounding wire** saves time when connecting the circuit grounding wire to the box, as explained in the next section on installing boxes (Figure 67).

The **grounding clip,** seen in Figure 68, is another item that may be used to connect the circuit grounding wire to the box.

The **cable clamp** is tightened against the cable near where it enters the box. Several other types are available (Figure 69). All are held in place by screws except the "**quick clamp**" which is already under tension (Figure 70).

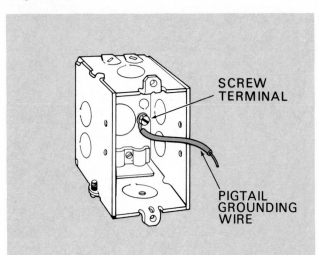

FIGURE 67. Grounding wire screw terminal installed in the tapped hole.

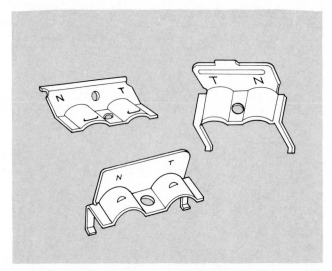

FIGURE 69. The cable clamp grips and holds the cable in the box when the screw is tightened.

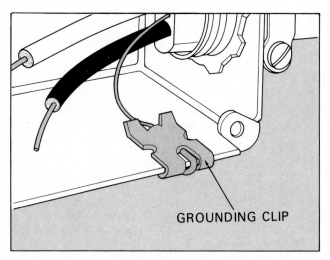

FIGURE 68. A grounding clip may be used to connect the grounding wire to the box.

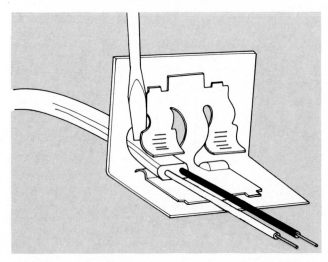

FIGURE 70. A quick-clamp type cable holder does not require a screw.

A **cable connector** may be installed where the knockout opening is used to bring in cable (Figure 71). Cable clamps inside the box are not required where connectors of this type are installed.

(2) Sizes of Device Boxes. The dimensions for device (switch) boxes are as follows: The sides are 7.6 cm (3 inches) high and the width 5.1 cm (2 inches). They are available with depths from front to rear of 3.8 to 8.9 cm (1½ to 3½ inches). More of the larger sizes such as 7.6 × 5.1 × 5.1 cm (3 × 2 × 2 inches) and 7.6 × 5.1 × 8.9 cm (3 × 2 × 3½ inches) are used (Figure 72).

Switch boxes may be purchased in sizes to accommodate two or more switches or receptacles side by side. Or you may make your own "gang boxes" as they are called. Remove one side on each box and put them together to provide space for two devices. You may gang as many as you need using the same procedure (Figure 73). To mount gang boxes, use metal or wood strips between studs to hold them in place (Figure 74).

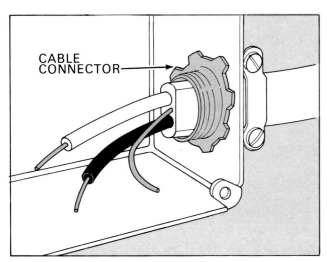

FIGURE 71. A cable connector installed in a knockout opening to hold the cable in place.

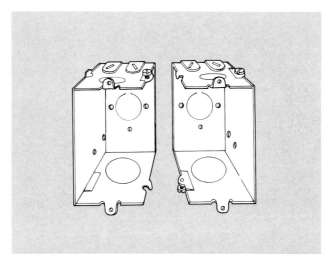

FIGURE 73. Two or more metal boxes may be fastened (ganged) together by removing one side on each box.

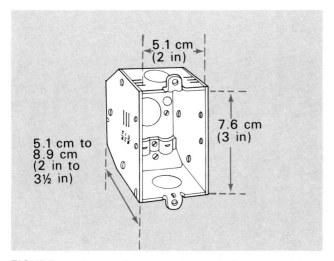

FIGURE 72. A large size switch box used in house wiring.

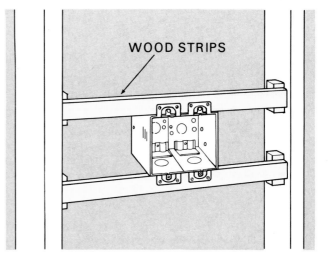

FIGURE 74. Ganged boxes may be mounted on metal or wood strips between studs to hold them in place.

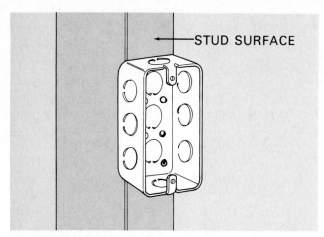

FIGURE 75. Handy boxes may be used for surface mounting switches or receptacles in such places as garages, basements and utility buildings.

In garages, basements and utility buildings, you may need a switch or receptacle that can be **surface mounted.** It is best to use a box known as a **handy box** for such locations (Figure 75). The handy box is a little larger than the standard switch box. Sizes are from 10.2 × 5.4 × 3.8 cm (4 × 2⅛ × 1½ inches) to 10.2 × 5.4 × 5.4 cm (4 × 2⅛ × 2⅛ inches). Device and handy box **covers** are available for use with either switches or receptacles (Figure 76).

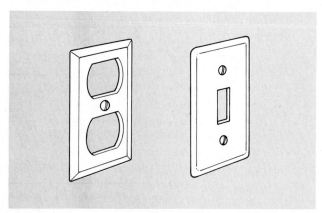

FIGURE 76. Covers available for use on handy boxes to mount switches or receptacles.

b. Types and Sizes of Outlet Boxes

Types and sizes of outlet boxes are discussed as follows:

(1) TYPES OF OUTLET BOXES. Outlet boxes may be octagonal or square (Figure 77). The octagon-shaped box is the common choice for overhead lighting fixtures and other locations requiring more room than a switch box provides. They are available with mounting brackets similar to those used on switch boxes. They may also be designed with the same type of knockouts, clamps and other fittings (Figure 78).

Switches and receptacles may be mounted in the octagon and square boxes if they are fitted with the correct type of cover (Figure 79). These larger boxes provide room for more wiring inside.

(2) SIZES OF OUTLET BOXES. Sizes of outlet boxes and device boxes are shown in Table IV. If you find

FIGURE 77. Outlet boxes may be octagonal or square.

FIGURE 78. Octagonal boxes are often used for ceiling light fixtures.They may have brackets and fittings similar to those on switch boxes.

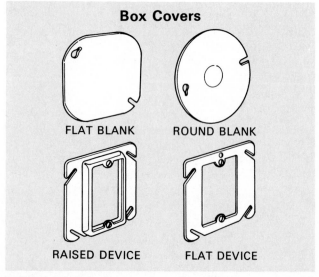

FIGURE 79. Types of outlet box covers.

TABLE IVA. MAXIMUM NUMBER OF CONDUCTORS TO BE INSTALLED IN METAL BOXES
[Customary Dimensions excerpted from Code Table 370–16(a)]

Box Dimensions, Inches Trade Size or Type	Min. Cu. In. Cap.	Maximum Number of Conductors						
		No. 18	No. 16	No. 14	No. 12	No. 10	No. 8	No. 6
4 x 1¼ Round or Octagonal	12.5	8	7	6	5	5	4	2
4 x 1½ Round or Octagonal	15.5	10	8	7	6	6	5	3
4 x 2⅛ Round or Octagonal	21.5	14	12	10	9	8	7	4
4 x 1¼ Square	18.0	12	10	9	8	7	6	3
4 x 1½ Square	21.0	14	12	10	9	8	7	4
4 x 2⅛ Square	30.3	20	17	15	13	12	10	6
4¹¹⁄₁₆ x 1¼ Square	25.5	17	14	12	11	10	8	5
4¹¹⁄₁₆ x 1½ Square	29.5	19	16	14	13	11	9	5
4¹¹⁄₁₆ x 2⅛ Square	42.0	28	24	21	18	16	14	8
3 x 2 x 1½ Device	7.5	5	4	3	3	3	2	1
3 x 2 x 2 Device	10.0	6	5	5	4	4	3	2
3 x 2 x 2¼ Device	10.5	7	6	5	4	4	3	2
3 x 2 x 2½ Device	12.5	8	7	6	5	5	4	2
3 x 2 x 2¾ Device	14.0	9	8	7	6	5	4	2
3 x 2 x 3½ Device	18.0	12	10	9	8	7	6	3
4 x 2⅛ x 1½ Device	10.3	6	5	5	4	4	3	2
4 x 2⅛ x 1⅞ Device	13.0	8	7	6	5	5	4	2
4 x 2⅛ x 2⅛ Device	14.5	9	8	7	6	5	4	2
3¾ x 2 x 2½ Masonry Box/Gang	14.0	9	8	7	6	5	4	2
3¾ x 2 x 3½ Masonry Box/Gang	21.0	14	12	10	9	8	7	4
FS–Minimum Internal Depth 1¾ Single Cover/Gang	13.5	9	7	6	6	5	4	2
FD–Minimum Internal Depth 2⅜ Single Cover/Gang	18.0	12	10	9	8	7	6	3
FS–Minimum Internal Depth 1¾ Multiple Cover/Gang	18.0	12	10	9	8	7	6	3
FD–Minimum Internal Depth 2⅜ Multiple Cover/Gang	24.0	16	13	12	10	9	8	4

For SI units: one cubic inch = 16.4 cm³.

during installation that you need more space than is available in a standard size box, you can add an extension ring to increase the capacity (Figure 80). Extension rings are also used to bring the edge of the box out to the surface when plaster or tile edges are beyond the box opening.

Table IV gives the number of wires the Code allows to be installed in each box where there are no fittings or devices such as fixture studs, cable clamps, switches and receptacles. The number of wires allowed depends on the size of the box and the size of the wires going into the box. However, most boxes have at least some number of cable clamps, studs and other items, in addition to the wires. The number of wires allowed in a box must be reduced to make room for these extra items.

The reductions required for items other than conductors can make a big difference in a size of the box needed. Detailed rules are given in the next section.

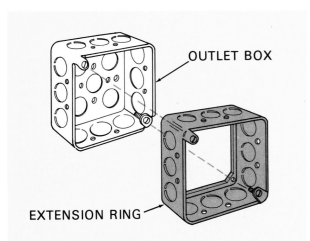

FIGURE 80. An extension ring added to a box to increase capacity or to extend box to flush surface.

TABLE IVB. VOLUME REQUIRED PER CONDUCTOR*
[Code Table 370–16(b)]

Size of Conductor	Free Space Within Box for Each Conductor
No. 18	1.5 cubic inches
No. 16	1.75 cubic inches
No. 14	2. cubic inches
No. 12	2.25 cubic inches
No. 10	2.5 cubic inches
No. 8	3. cubic inches
No. 6	5. cubic inches

*Note that wire sizes No. 18 and No. 16 have been added to 370-6(b). These sizes must now be included in calculations to determine the maximum number of conductors allowed in a box. Also note that where conductors of more than one size enter a box, the largest conductor size must be used when calculating the number of deductions from the maximum number.

2. NUMBER OF CONDUCTORS ALLOWED IN BOXES
(Code Section 370-16a)

The Code requires boxes to be large enough to provide free space for all wires brought into the box. Table IV shows the maximum number of wires by sizes that can be brought into metal boxes of various dimensions, according to the Code.

The number of wires allowed per box according to Table IV must be reduced by one for the following contained in a box:

—For one or more studs or cable clamps or hickeys, reduce by one for each type of device. If one of each is included, reduce by three (Figure 81).

—For each strap containing one or more devices such as a switch or receptacle, reduce by two (Figure 81).

—For one or more grounding wires entering the box, reduce by one.

—For each conductor originating outside the box and ending inside the box, reduce by one.

—For each conductor running through the box, reduce by one.

If no part of a conductor leaves the box, no reduction is required for that conductor, for example, a pigtail ground wire (Figure 82).

The importance of the Code restrictions may be seen in the following example. Suppose you want to install a wall-mounted light in a bathroom as shown in the house plan, using No. 12 wire. Following Code requirements in the table, you must reduce the number of conductors allowed in the box as follows (Figure 83):

—For one stud, reduce number allowed by . . . . 1 (a reduction of one, whether one or two cable clamps)

—For two cable clamps, reduce number allowed by 1

—For one strap, reduce number allowed by . . . 2

—For one or several grounding wires, reduce number allowed by 1

—**Total reduction** **5**

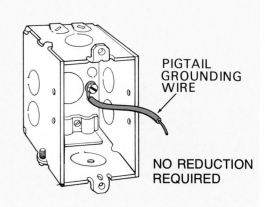

FIGURE 82. A pigtail ground wire is confined to the box, therefore, it is not counted against the maximum number of conductors to be allowed in a box.

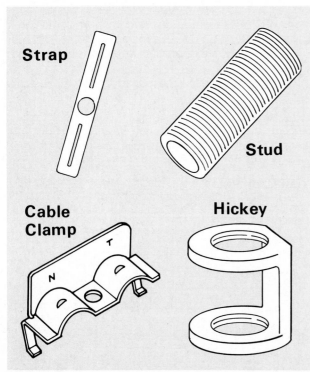

FIGURE 81. Strap, stud, cable clamp and hickey. Reduce the number of conductors allowed in a box as per NEC®(370-16(a).

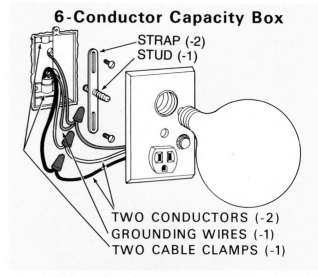

FIGURE 83. Box must be large enough for 7 conductors for this 2-conductor light outlet.

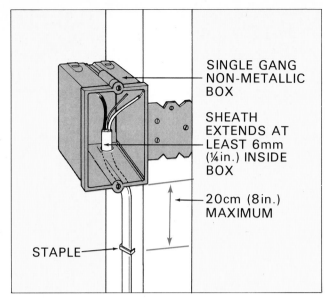

FIGURE 84. Single gang nonmetallic box, without cable clamp. The cable is not required to be secured in a nonmetallic box (Code Section 370-7 [c] Ex. 1).

To allow space for the two conductors required for the light outlet, you must have a device box large enough for seven conductors, according to the table.

The minimum size of the device box for No. 12 wire is 7.6 x 5.1 x 6.4 cm (3 x 2 x 3½ inches) (Table IV). For No. 14 wire, the minimum size device box is 7.6 x 5.1 x 5.7 cm (3 x 2 x 2¾ inches).

Using nonmetallic boxes will allow more wires to be installed than in a metal box of the same size because the clamps may be omitted.

Where nonmetallic sheathed cable is installed in **nonmetallic boxes no larger than 5.7 cm × 10.2 cm** (2¼ in x 4 in) and mounted in walls, the Code does not require the cable to be clamped to the box, provided the cable is supported within 20 cm (8 inches) of the box and where the sheath extends into the box at least .6 cm (¼-inch) (Figure 84).

3. INSTALLING SWITCH BOXES AND HANDY BOXES

Code Section 370-20 requires that switch boxes and other outlet boxes be mounted flush with the edge of the wall surface on combustible materials such as wood paneling (Figure 85). If the surface of the drywall plaster or gypsum board is damaged around the opening for the box, it must be repaired so there are no gaps or openings more than .3 cm (⅛-inch) at the edge of the box, according to Section 370-21 of the Code. On plaster, concrete or gypsum wallboard, the box may be recessed not more than 6mm (¼ inch) in back of the finished surface. Many boxes are manufactured with depth gage marks on the side. The marks save time

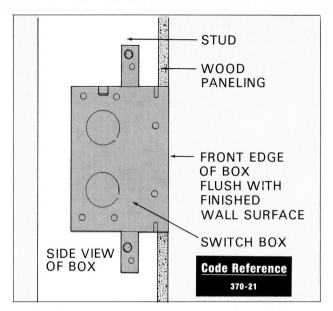

FIGURE 85. Boxes must be mounted flush with the edge of finished wall surface on combustible materials.

by eliminating the need for measuring at each mounting. The lines on some boxes show the depth of each mark, for example, 1, 1.3, and 1.6 cm (⅜, ½, and ⅝ inches) (Figure 86). Others have the lines without the figures.

Installing switch boxes and handy boxes is discussed under the following headings:

a. Mounting Switch Boxes on Studs.
b. Mounting Switch Boxes in Old Buildings.
c. Installing Switch Boxes Between Studs.
d. Installing Handy Boxes.

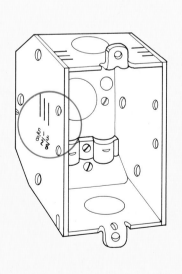

FIGURE 86. Box with depth gage marks on the side, eliminating the need for measuring at each mounting.

a. Mounting Switch Boxes on Studs

The most common method of installing a switch box is to nail it to the stud. Keep in mind that switches and receptacles are both commonly mounted in switch boxes.

Proceed as follows to mount switch boxes on studs:

1. *Remove knockouts or pryouts.*

   It is best to remove **knockouts** and **pryouts** before mounting the box. Remove only those needed. Any knockout and pryout openings not used must be covered with knockout seals. Determine where the openings are needed and loosen the knockout plug with a sharp blow. Remove with pliers. To remove pryouts, insert a screwdriver in the slot, twist and pry.

2. *Determine mounting depth.*

   (1) *Check depth gage markings on side of box and select line that corresponds to needed depth.*

      For example, if wallboard or plaster is 1.3 cm (½ inch) thick, align the box on that mark (Figure 87).

   (2) *If the box has a side-mounted bracket, be sure to buy boxes with brackets that are backset to the correct depth for flush mounting with finished wall surface (Figure 88).*

   (3) *If no marks are present, measure the depth with a ruler or gage.*
      Mark the line (Figure 89).

3. *Nail box to stud.*

   (1) *Drive two nails through side holes in the box (Figure 90) or through bracket top and bottom (Figure 91).*

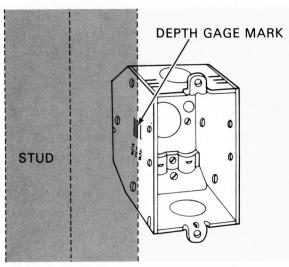

FIGURE 87. Marker on the box aligned with front edge of the stud.

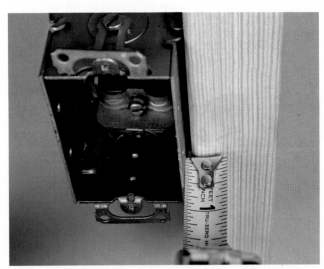

FIGURE 89. Marking mounting depth for wallboard thickness.

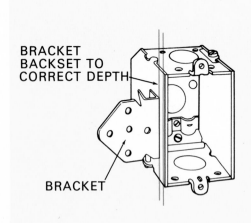

FIGURE 88. Box with side-mounted brackets backset for wallboard thickness.

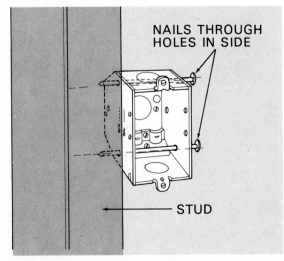

FIGURE 90. Nails driven through holes on sides of box.

No. 10d galvanized nails are preferred by many. The galvanized coating gives them added holding strength.

(2) *To mount side-bracket box, hold bracket against stud and tap bracket points into place.*

Drive nails into holes in bracket (Figure 92).

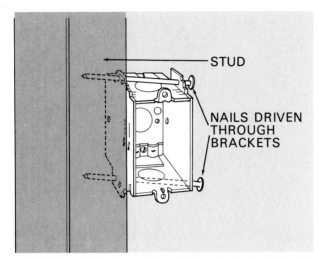

FIGURE 91. Nails driven through brackets on top and bottom of box.

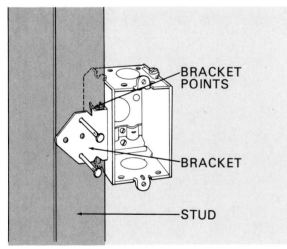

FIGURE 92. Side-bracket box mounted by tapping bracket points into stud, then driving nail through the bracket holes.

b. Installing Boxes in Old Buildings

To mount a new box in an old building, a special type of box or supports may be required. Proceed as follows:

1. *Use a template or outline of your box to mark lines to be cut for opening in the wall.*

2. *Drill holes at opposite corners to start the cut.*
3. *Finish cut with a keyhole saw or other type of saw (Figure 93A).*
4. *Insert a special box with cable installed into the opening.*

For a box with side-bracket supports, tighten the screws to bring the brackets up snug against wall (Figure 93B). Several other types of supports are available. Follow directions included with your box to install.

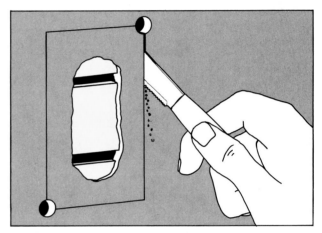

FIGURE 93A. Keyhole saw or hacksaw blade will cut hole for box. page 45

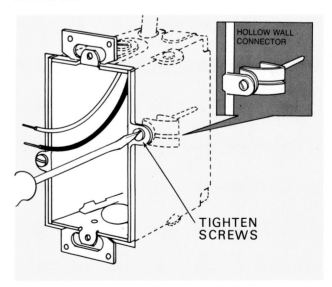

FIGURE 93B. Tighten screw on each side of box to pull the brackets up tight against the inside wall.

c. Installing Switch Boxes Between Studs

You will sometimes find that you need to mount a switch box or gang boxes at a point that falls between two studs. When that happens, you must mount a support between the studs (Figure 94). It may be a steel support strip with mounting holes punched or two wooden strips at least 2.5 cm (1 inch) thick by 5.1 cm (2 inches) wide (Figure 95). Proceed as follows:

1. *Mark location for each strip.*
   Use the switch box as a spacer.
2. *Nail strips in position between studs.*
3. *Locate one or more boxes at the desired point on the support.*
4. *Fasten box to strip with metal screws or wood screws.*

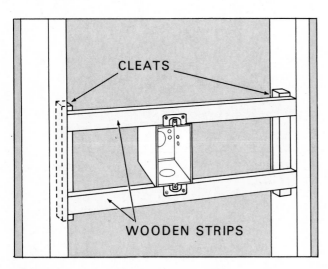

FIGURE 94. **Box mounted between studs.**

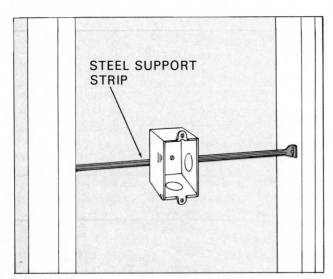

FIGURE 95. **Box mounted between studs on steel support strip.**

d. Installing Handy Boxes

Handy boxes are surface mounted in such places as garages, basements, utility buildings, and agricultural buildings. Proceed as follows:

1. *Remove knockouts needed.*
2. *Place handy box in position as marked.*
3. *Nail two or more 2.5 cm (1 inch) nails through holes located in bottom of box or use screws for mounting (Figure 96).*
4. *For concrete blocks, drill holes where required and place inserts for screws (Figure 97).*
   Mount box and fasten with screws.

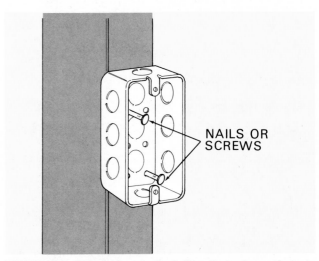

FIGURE 96. **Handy box nailed directly to face of stud.**

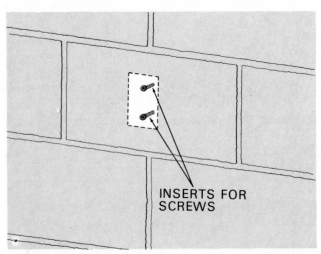

FIGURE 97. **Holes drilled in concrete blocks and inserts placed for screws.**

4. INSTALLING OUTLET BOXES

You now have procedures for installing a switch box or handy box at every point in the house where a switch or receptacle is to be installed. However, one 120-volt receptacle in the standard house plan must have a larger box to accommodate more wiring. Refer to Figure 52. Note the outlet indicated by the arrow as the starting point for circuit No. 5. The cable enters the box from the SEP. Wiring must then be installed for a run in two directions. One run is to the receptacle near the dining room door. Another run is to the rest of the circuit. A switch box would not provide enough space for the required wiring. A square outlet box, size 10.2 × 10.2 × 5.4 cm (4 × 4 × 2⅛ inches) or similar size should be installed to allow room for the wiring required and for the duplex receptacle. Attach special-made cover for mounting the receptacle (Figure 98).

Outlet boxes may be **octagonal** in shape or square. Procedures for installation are the same for both types. The octagonal box is the usual choice for overhead lighting outlets. The square boxes are used where more space is needed. Both types are available with brackets similar to those used for switch boxes. Others are mounted on bar hangers.

Installing outlet boxes is discussed under the following headings:

a. Installing Outlet Boxes with Mounting Brackets.
b. Installing Outlet Boxes with Bar Hangers.

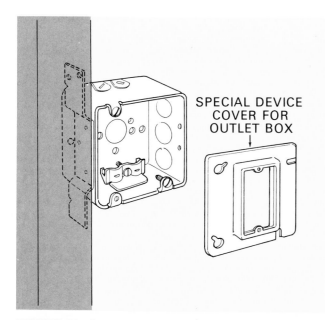

FIGURE 98. Special cover for receptacle mounting in large box.

SPECIAL DEVICE COVER FOR OUTLET BOX

a. Installing Outlet Boxes with Mounting Brackets

Proceed as follows:

1. *Remove knockouts as needed.*
2. *Line up bracket with stud.*
   Use brackets that are set at the correct depth for the wall covering.
3. *Tap in the bracket point on the bracket to hold the box in position.*
4. *Nail the bracket to the stud using two or more nails (Figure 99).*

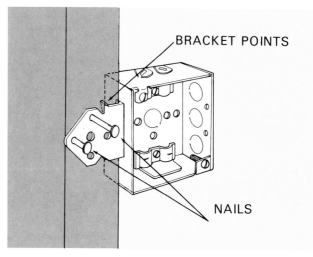

BRACKET POINTS

NAILS

FIGURE 99. Bracket of outlet box nailed to the stud.

b. Installing Outlet Boxes with Bar Hangers

When used for mounting overhead fixtures, the octagonal box may be installed several different ways. Most often it will be mounted with a metal bar hanger attached between the joists. The bar and hanger may be **bought separately** and assembled on the job or you may buy it assembled and ready to hang. The bar hanger may be (1) **adjustable,** or (2) **solid.**

(1) ADJUSTABLE BAR HANGERS. Proceed as follows to attach and install the **adjustable bar hangers,** if not assembled.

1. *Remove center knockout from box (Figure 100).*
2. *Remove clamping screw and fitting from bar (Figure 101).*
3. *Place box on clamp, insert screw and fitting (Figure 102).*

4. *Tighten screw loosely against fitting (Figure 103).*

   The box must move freely to adjust the position after hanging.
5. *Mount hanger bar between joists.*

   Adjust to desired depth of recess. Adjust bar to fit between joists and nail at each end (Figure 104).

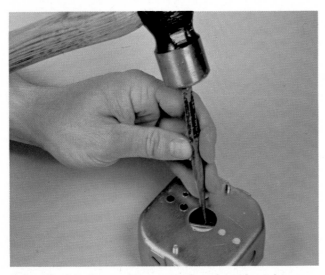

FIGURE 100. Removing center knockout from box.

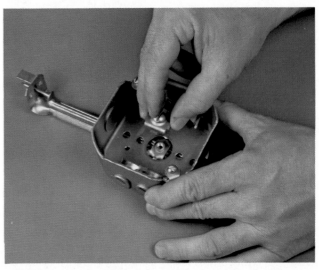

FIGURE 102. Box placed on the clamp and screw and fitting inserted.

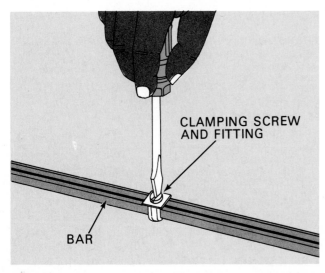

FIGURE 101. Removing clamping screw and fitting from bar.

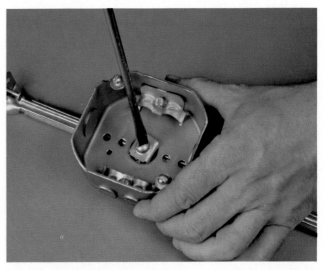

FIGURE 103. Screw tightened loosely to allow adjustment after hanging.

FIGURE 104. Hanger bar placed in position and nailed at each end between joists.

6. *Move box to desired position on bar and tighten screw firmly against fitting.*

If adjustable bar and hanger are **already assembled,** follow only steps 5 and 6 above.

Several types of pre-assembled and pre-wired **recessed ceiling fixtures** are available. The mounting procedure for adjustable hanger bars on these units is the same as for octagon box hangers except that there are two hanger bars to adjust (Figure 105). Procedures for steps 5 and 6 will vary depending on how the fixture is fastened to the hanger bar. Tighten screws at side of each bar.

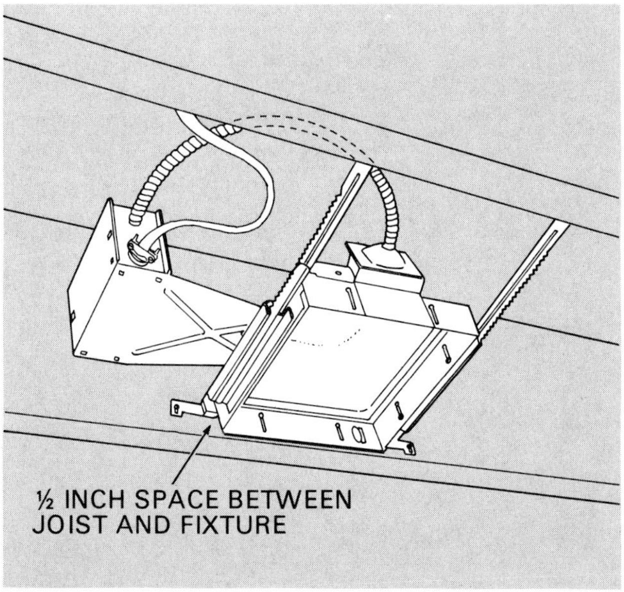

(2) SOLID BAR HANGERS may be straight or offset (Figure 106). They may be ordered in overall lengths of 46, 61 and 76 cm (18, 24 and 30 inches). Assembly procedures are the same as for adjustable bars if not assembled. Proceed as follows:

1. *Follow steps (1) through (4) as for unassembled adjustable bar hangers.*
2. *To hang the bar, either straight or offset, place the bar in position across the joist edges and nail each end.*

   Cut notches if needed.
3. *Move the box to the desired position and tighten the screw against the fitting (Figure 107).*
4. *To hang pre-assembled fixture, place fixture in position between joists and nail both bars at each end (Figure 108).*

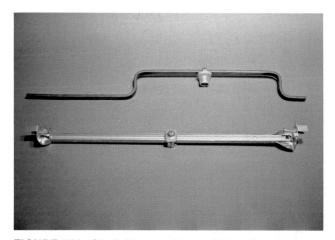

FIGURE 106. Straight and offset solid bar hangers.

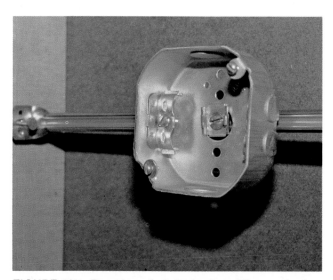

FIGURE 105. Hanger bar adjusted on each side of preassembled, recessed ceiling fixture.

½ INCH SPACE BETWEEN JOIST AND FIXTURE

FIGURE 107. Box placed at desired position on the bar and screw tightened.

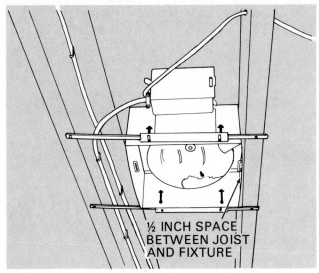

FIGURE 108. Preassembled fixture in position.

E. Providing Access for Nonmetallic Sheathed Cable

Most residences are wired with nonmetallic sheathed cable. It is one or more wires wrapped in insulation.

Providing access is another way of saying that you must provide a path for a circuit of nonmetallic sheathed cable to follow. From the SEP to the outlets, cable must be protected in some way from possible damage after it is installed.

Cable protection may be provided by drilling holes or cutting notches in studs, joists and rafters. In attics and unfinished basements, cable may be placed on running boards or between guard strips.

Providing access for nonmetallic sheathed cable is discussed as follows:
1. Drilling Holes in Framework.
2. Cutting Notches in Framework.
3. Installing Guard Strips.
4. Installing Running Boards.

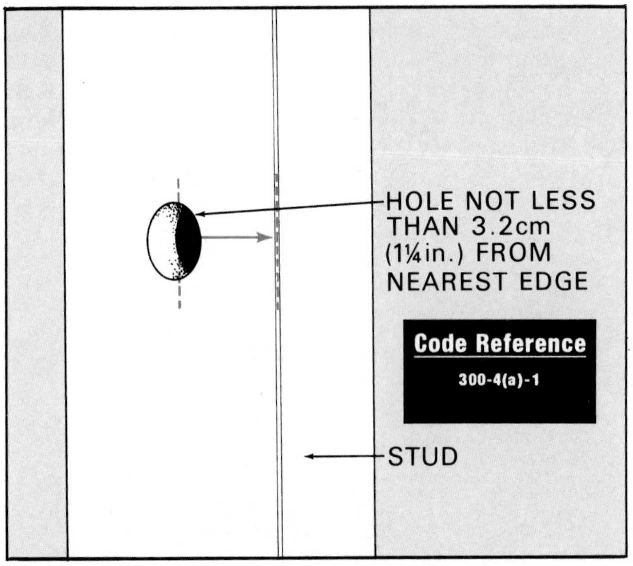

HOLE NOT LESS THAN 3.2cm (1¼ in.) FROM NEAREST EDGE

Code Reference
300-4(a)-1

STUD

FIGURE 109. Hole in stud for supporting cable.

1. DRILLING HOLES IN FRAMEWORK (Code Section 300-4(a)1)

Holes drilled for cable installation must be placed at the approximate center of the framing member. The edge of holes in studs must be not less than 3.2 cm (1¼ inches) from the nearest edge of the stud (Figure 109). If less than 3.2 cm (1¼ inches), the cable must be protected from nails by either a bushing or a steel plate at least 1.6 mm (1/16-inch) thick (Figure 110). Holes in the center are at neutral axis and do not weaken the structure in horizontal framing members.

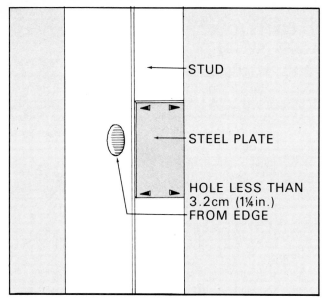

FIGURE 110. Steel plate for protecting cable from nails if hole less than 3.2 cm (1¼ inches) from nearest edge of stud.

Having the right equipment for the job makes the drilling go faster. The hand operated brace and bit has largely been replaced by power drills for drilling holes in framing. A power drill equipped with a boring bit cuts holes fast. A 1.6 cm (⅝-inch)-bit will fit most requirements for cable openings in houses. The flat bits are available in sizes from 6 mm (¼ inch) up (Figure 111).

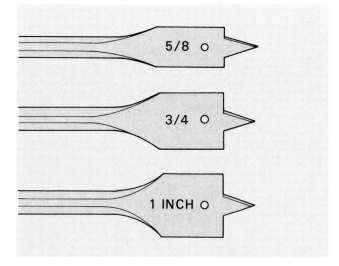

FIGURE 111. Bits for cutting holes in framing.

With switch boxes, receptacle boxes and outlet boxes in place, you may drill holes between them to provide access for the cable. Proceed as follows:

1. *Check circuit routes to see where drilling is required.*

    Refer to Figure 41 for small appliance circuit No. 1. This circuit has 6 outlets. Figure 42 shows how the circuit is routed and, therefore, where holes are needed in studs above the kitchen counter. Figure 43 shows a continuation of the same circuit with holes drilled in a dining room wall.

2. *Drill holes for small appliance circuits.*

    Locate as shown in Figure 42. Line up the circuit route by sight. Drill a hole in the center of each stud, or at least 3.2 cm (1¼ inches) from the nearest edge. Keep all holes in line between the boxes. Holes that are out of line are harder to pull cable through. They also require more wire. Where it is necessary to change elevation of holes, make gradual changes without sharp turns (Figure 112).

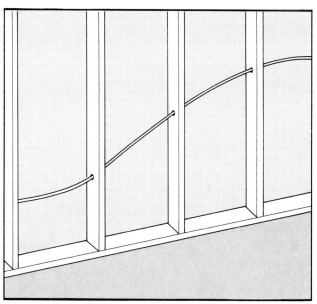

FIGURE 112. Holes drilled for changes in elevation to avoid sharp turns.

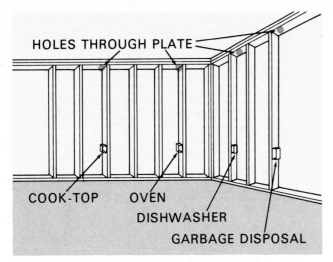

FIGURE 113. Holes drilled in top plate for individual overhead circuits to reach outlets on studs.

3. *Drill a hole in the plate for each individual kitchen circuit directly above the outlet box (Figure 113).*

   Note the location of the four individual kitchen circuits, the garbage disposer, dishwasher, oven and cooking top. The oven and cooking top will each have an individual circuit that runs overhead from the SEP. The dishwasher and garbage disposer may be combined on one circuit if desired. If so, drill holes between the outlets for the connecting cable.

4. *Drill required holes for switches.*

   Check the location of each switch outlet in the kitchen and dining room and drill a hole in the plate directly above each one for overhead wiring (Figure 114).

5. *Drill holes for general purpose circuits (Figure 44).*

   Follow procedures as for small appliance circuits.

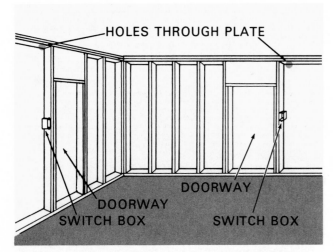

FIGURE 114. Holes drilled in top plate for switches.

52

2. CUTTING NOTCHES IN FRAMEWORK

The Code permits the installation of cable in notches cut into the framing if it can be done without weakening the building structure. However, you must protect the cable from screws and nails with a steel plate at least 1.6 mm (1/16-inch) thick at those points (Figure 115).

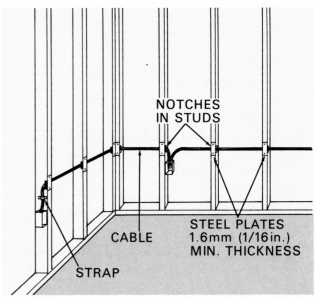

FIGURE 115. Notches in studs for cable route covered by steel plates.

Recent developments in housing design may bring about greater use of notching or drilling as a means of improving the efficiency of insulation. Many house plans designed for maximum energy conservation now call for 5.1 × 15.2 cm (2″ × 6″) studs in the outside walls. The wider studs permit installation of six inches of insulation. Wiring installed in holes through the center of studs would interfere with installing batt or roll-type insulation and with its efficiency after installation. **Notching or drilling such studs for cable at the bottom** would eliminate much of the problem (Figure 116). A similar problem is present with 5.1 × 10.2 cm (2″ × 4″) studs when insulation is installed. A portable saw or drill should be used for speedier notching or drilling.

Two methods for notching or drilling are given as follows:

   a. Before Stud Installation.
   b. After Stud Installation.

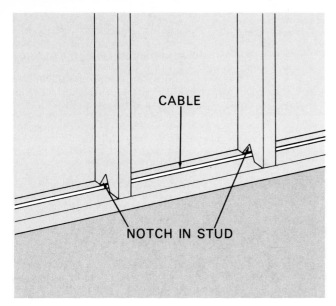

FIGURE 116. Studs notched at the bottom for cable installation permit more efficient use of insulation between studs.

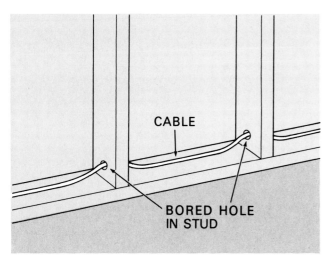

FIGURE 117. Holes drilled at bottom of studs.

a. Before Stud Installation

Proceed as follows:

1. *Before erecting each stud, cut a V-shaped notch or drill a hole in one end.*

   The mouth of the cut can be 2.5 cm (1 inch) wide on 5.1 × 10.2 cm (2″ × 4″) studs and much wider on 5.1 × 15.2 cm (2″ × 6″) studs. Taper the cut to the center of the stud (Figure 116).

   For drilling, a ⅝-inch hole is satisfactory (Figure 117).

2. *After studs are erected, insert the cable through the opening.*

   The cable rests on the sole plate out of the way of insulation.

b. After Stud Installation

Proceed as follows to notch or drill installed studs:

1. *Line up the notches by sight.*
2. *Cut notch as wide and deep as required for a loose fit of cable.*
3. *Use a wood chisel to remove the chip (Figure 118).*
4. *Repeat for all remaining cuts.*
5. *Install an individual steel plate ¹/₁₆-inch thick at each cut (Figure 119).*
6. *If using a drill, make the hole as near the bottom of the stud as possible.*

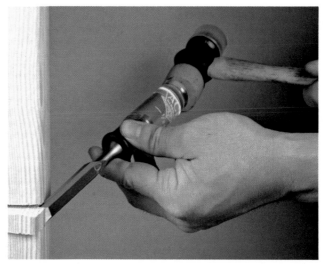

FIGURE 118. Removing the chip with a wood chisel.

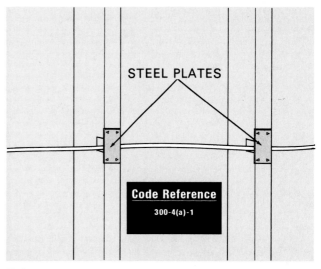

Code Reference
300-4(a)-1

FIGURE 119. Steel plates installed over the notches.

3. INSTALLING GUARD STRIPS

If an attic is not easily accessible and not floored, cable must be protected only within 1.8 m (six feet) of the attic entrance. In other parts of the attic, cable may be run across the top of the ceiling joists. If the attic is easily accessible by stairs or permanent ladder, a cable run across the face of floor joists, rafters or studding must be protected either by running it through holes drilled in ceiling joists or by providing guard strips that are at least as high as the cable (Figure 120).

Proceed as follows to install guard strips:
1. *Determine route to be followed by cable run.*
2. *Nail strips approximately 1.3 cm (½-inch) apart for the length of the run.*
3. *The cable is installed between strips and fastened to framing.*

   Cables that are run parallel to joists or rafters must be installed so that the nearest outside surface of the cable is not less than 1¼ inches from the nearest edge of the framing member. If the cable is closer, the cable must be protected by a steel plate or sleeve ¹⁄₁₆ of an inch thick.

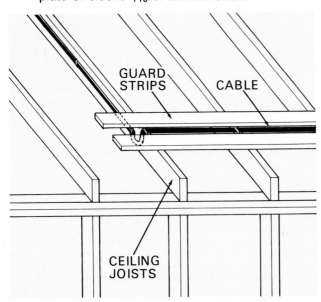

FIGURE 120. Guard strips installed in the attic.

4. INSTALLING RUNNING BOARDS

When cable is run across joists in unfinished basements, you are permitted to fasten cables of two No. 6 or three No. 8 conductors or larger, directly to the lower edges of the joists (Figure 121). Smaller cables must be run either through holes drilled in joists or on running boards (Figure 122). A running board may be 2.5 × 10.2 cm (1″ × 4″) or other size board, placed end to end over the cable run. Cable is stapled to the board.

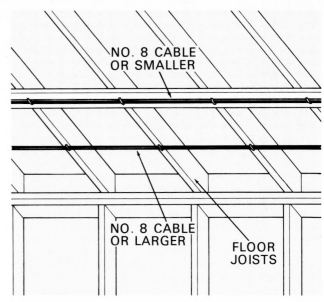

FIGURE 121. In unfinished basements, cable size No. 8 or larger may be fastened directly to lower edge of the joists. Smaller cable must be fastened to running boards.

Proceed as follows to install running boards.
1. *Determine route to be followed by the cable.*
2. *Nail running board in place (Figure 121).*

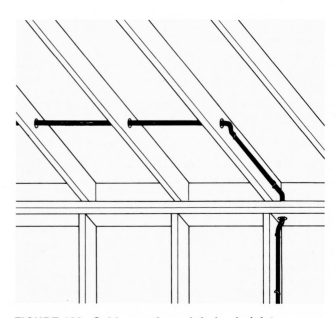

FIGURE 122. Cable run through holes in joists.

F. Pulling Cable for Circuits

Your next step is to pull the cable for each circuit. Pulling cable for circuits is discussed under the following headings:
1. Types and Characteristics of Cables.
2. Removing Cable from the Roll.
3. Pulling the Cable.

1. TYPES AND CHARACTERISTICS OF CABLES

Types and characteristics of conductors and cable are discussed as follows:
a. Types of Cable.
b. Characteristics of Cable.

a. Types of Cable

Types of cable are discussed under the following headings:
(1) Types of Wire.
(2) Characteristics of Wire and Insulation.
(3) Characteristics of Cable.
(4) Nonmetallic Sheathed Cable.
(5) Cable Identification.
(6) Cable Ampacity.

(1) TYPES OF WIRE are discussed as follows:
— Copper Wire.
— Aluminum Wire.

More **copper wire** is used in house wiring than any other type (Figure 123). It is a better conductor of electrical current than **aluminum wire** and gives fewer problems after it is installed. But in larger wire sizes, copper is considerably more expensive than aluminum or copper-clad aluminum (aluminum wire coated with copper). On installations requiring wire sizes **larger than No. 8,** such as feeder lines and service entrances, aluminum is gaining in popularity because it is less expensive.

Number 14 is the smallest size copper wire the Code permits for house wiring. However, many local codes require that nothing smaller than No. 12 copper be used.

Because aluminum is a poorer conductor of electrical current than copper, aluminum wire must be larger than copper to carry the same current load. Generally, aluminum wire must be one trade size larger than copper to be equivalent. In an installation where No. 8 copper is satisfactory, a No. 6 aluminum wire would be required to carry the current load (Figure 124). In the same manner, No. 10 aluminum could be substituted for No. 12 copper wire.

NOTE: Size increments come from ampacity tables in the Code. For example, the difference between AWG sizes 6 and 8 or between 8 and 10 is generally considered to be one size. The Code ratings are listed only for commonly manufactured sizes.

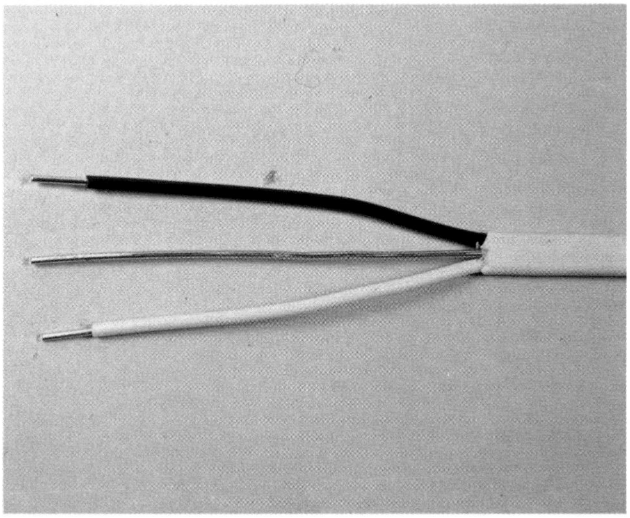

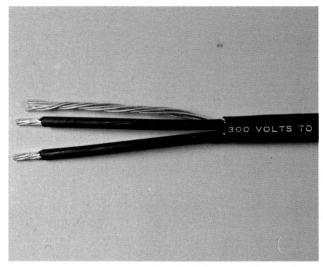

FIGURE 123. Copper wire is most often used in house wiring.

FIGURE 124. In an installation where No. 8 AWG copper wire is required, aluminum wire must be No. 6 AWG to conduct the current load.

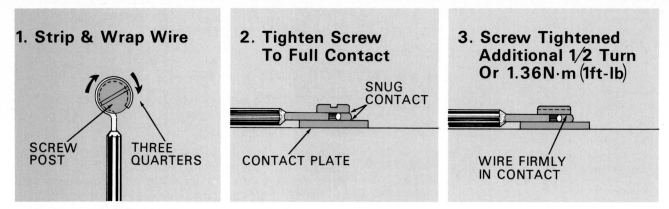

| 1. Strip & Wrap Wire | 2. Tighten Screw To Full Contact | 3. Screw Tightened Additional 1/2 Turn Or 1.36N·m (1ft-lb) |

SCREW POST   THREE QUARTERS

SNUG CONTACT

CONTACT PLATE

WIRE FIRMLY IN CONTACT

FIGURE 125. Steps required to connect wire to a terminal screw correctly.

All references to wire size refer to copper wire unless aluminum is indicated. If aluminum wire is to be used instead, remember that it must be one size larger.

Terminations of smaller sizes of **aluminum wire**, such as those used for interior wiring of homes, have caused problems in many installations in the past. One reason is that older types of switches, receptacles and other devices not made especially for use with aluminum proved unsatisfactory. Expansion and contraction at terminals caused connections to become loose and begin arcing.

Another cause of problems with copper or aluminum wiring is poor workmanship. Installers must be especially careful when making connections. A terminal screw that is not tight, or a wire loop overlapped under a terminal screw, is likely to cause problems.

Steps for properly connecting wire to a terminal screw are shown in Figure 125.

Newer **receptacles** and **switches** made for use with either copper or aluminum and rated 15- or 20-amp are marked CO/ALR. Always use devices carrying only

this mark with aluminum cable (Figure 126). **Equipment** rated 30 amperes and over with terminals listed for either copper or aluminum wire is labeled AL-CU.

Copper-clad aluminum wire is aluminum wire that has a copper coating (Figure 127). It may be used with all wiring devices currently listed by UL for use with copper, including the push-in backwired type. Ampacity ratings for copper-clad aluminum are the same as for uncoated aluminum.

(2) CHARACTERISTICS OF WIRE AND INSULATION. To understand the types of cable, you must become familiar with the types of conductors (wires) and insulation used to manufacture cable. Table 310–13 of the Code contains several pages of descriptive information on conductors used for general wiring. The information in Table V is an excerpt from the Code table which describes conductors used in house wiring. It lists trade names, types of wire, maximum operating temperature, types of locations where each may be used, the insulating material of each wire type and its characteristics, and the type of outer cover (sheath).

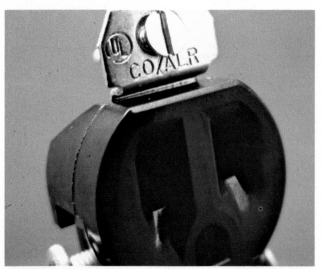

FIGURE 126. Connect aluminum wire to devices marked CO/ALR only.

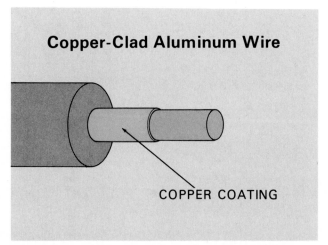

Copper-Clad Aluminum Wire

COPPER COATING

FIGURE 127. Copper-clad aluminum wire is aluminum wire that has a copper coating.

TABLE V. CONDUCTOR APPLICATION AND INSULATIONS
(See NEC® Section 310-13 for a complete cable guide.)

Trade Name	AWG Range	Applications			Insulation Thickness	Outer Jacket	Temperature
		Dry	Damp	Wet			
TW*** Moisture and heat resistant thermo-plastic	14 thru 2000 kcmils	X		X	30 to 125 mils	none	60 C 140 F
THHN Heat resistant thermo-plastic	14 thru 1000 kcmils	X	X		15 to 70 mils	nylon	90 C 197 F
THHW Moisture & heat resistant thermo-plastic	14 thru 1000 kcmils			X	40 to 110 mils	none	75 C 194 F
		X					90 C 194 F
THWN * ****** Moisture & heat resistant Thermo-plastic	14 thru 1000	X		X	15 to 70 mils	nylon	75 C 167 F
UF Underground feeder & BC cable	14 thru 4/0	See article 339 NEC®			60 to 95* mils	integral	60 C 140 F
							75 C 167 F
USE**** underground SE conductor or cables	12 thru 2000 kcmils	See article 338 NEC®			40 thru 125 mils Moisture** resistant	nonmetallic covering	75 C 167 F

* Includes integral jacket.

** Insulation thickness shall be permitted to be 80 mils for listed Type USE conductors that have been subjected to special investigations. The nonmetallic covering over individual rubber-covered conductors of aluminum-sheathed cable and of lead-sheathed or multiconductor cable shall not be required to be flame retardant. For Type MC cable, see Section 334-20. For nonmetallic-sheathed cable, see Section 336-25. For Type UF cable, see Section 339-1.

*** Insulation and outer coverings that meet the requirements of flame-retardant, limited smoke and are so listed shall be permitted to be designated limited smoke with the suffix /LS after the Code type designation.

**** Listed wire types designated with the suffix -2 such as RHW-2 shall be permitted to be used at a continuous 90° C operating temperature, wet or dry.

In the manufacture of nonmetallic-sheathed cable, two or more insulated conductors such as those in Table V are encased in an outer sheath to make a cable.

Some characteristics of insulation on conductors included in Table V are:

(a) Moisture and Heat Resistant Latex Rubber is for use in dry and wet locations.

(b) Moisture Resistant Thermoplastic* Type TW is suitable for both dry and wet locations. This is the type most commonly used for conductors in nonmetallic cable.

(c) Heat Resistant Thermoplastic* Type THHN withstands high temperatures and can be used for dry and wet locations. It is often used in conduit because the insulation is thinner. Thus more conductors can be installed in a given conduit.

(d) Thermoplastic* Types THW and THWN may both be used in dry and wet locations. Both are moisture and heat resistant thermoplastic* material.

(e) Type XHHW is a cross-linked synthetic polymer material that is moisture- and heat-resistant.

The thickness of the insulation on each type depends on the wire size.

(f) Type UF is used for direct burial of underground feeder circuits, such as between the residence and a separate utility building or garage. It cannot be buried in concrete.

(g) Type USE is for direct burial of underground service entrance, feeder circuits, and for indoor circuits.

(3) CHARACTERISTICS OF CABLE. Table V describes characteristics, insulations and conditions for use of individual wires. Most house wiring is done with **cable** made of the individual wires. Each type of cable also has its own characteristics and conditions for use. **The type of outer covering, or sheath, and the type and number of wires contained in the cable determine where and how the cable may be used.**

A cable containing two No. 12 wires is called a 12-2 cable. A cable having three No. 12 wires is a 12-3 cable. If cable also contains a grounding wire, it is called a 12-2w/g (with ground) (Figure 128) or a 12-3w/g (with ground).

The Code requires color coding of insulated wires in a cable. They are usually colored as follows:

—**Two wires**—one black, one white** (Figure 129).

—**Three wires**—one black, one white**, one red (Figure 130).

---

*Thermoplastic insulation may stiffen at temperatures colder than minus 10°C (plus 14°F), requiring care be exercised during installation at such temperatures. Thermoplastic insulation may also be deformed at normal temperatures where subjected to pressure, requiring care be exercised during installation and at points of support.

**A natural gray color may be used in place of white, but white is much more common.

Two-Wire Cable With Bare Grounding Wire

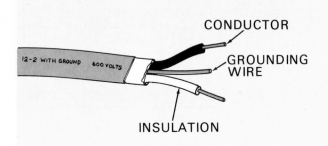

FIGURE 128. A 12-2 cable with ground wire.

Two-Wire Cable Without Grounding Wire

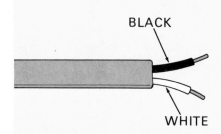

FIGURE 129. In a two-wire cable, the colors are usually one black, and one white.

Three-Wire Cable Without Grounding Wire

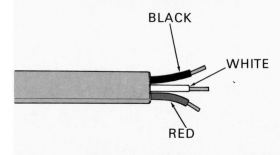

FIGURE 130. Wires in a three-wire cable are usually colored black, white and red.

1. DRILLING HOLES IN FRAMEWORK
(Code Section 300-4 [a] 1)

If the cable contains an insulated grounding wire, it is green or green with yellow stripes, according to Code Section 210-5 (b) (Figure 131). More often the grounding wire is bare (Figure 132).

(4) NONMETALLIC SHEATHED CABLE is often called Romex, which is the trade name of one manufacturer. The sheath on nonmetallic-sheathed cable is moisture-resistant, flame retardant, nonmetallic material. The sheath is usually made of plastic on cable used for house wiring.

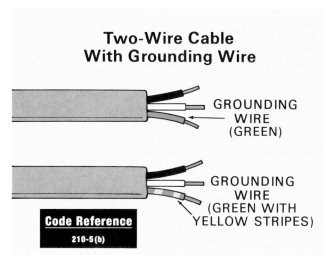

Two-Wire Cable With Grounding Wire

GROUNDING WIRE (GREEN)

GROUNDING WIRE (GREEN WITH YELLOW STRIPES)

Code Reference
210-5(b)

FIGURE 131. An insulated grounding wire in a cable is green or green with yellow stripes.

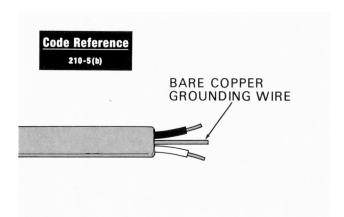

Code Reference
210-5(b)

BARE COPPER GROUNDING WIRE

FIGURE 132. The grounding wire in non-metallic sheathed cable is usually bare.

Nonmetallic-sheathed cable is available with either two or three insulated conductors. Also, it is available with an additional grounding wire which may or may not be insulated (Figure 133). Code Section 210-5 (a) should be referenced for additional information. Most cable installed in a house must have a grounding wire to meet local code requirements.

Two types of nonmetallic-sheathed cable are used in house wiring. Both types are manufactured in copper in sizes 14 through 2 (Figure 134). If the conductors are aluminum, or copper-clad aluminum, the available sizes are 12 through 2. They are described as follows:
(a) Type NM.
(b) Type NMC.

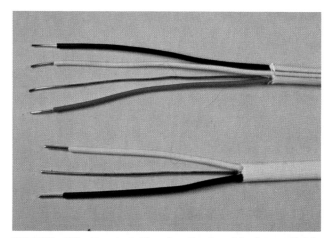

FIGURE 133. Nonmetallic sheathed cable is available with either two or three insulated conductors and with or without a bare grounding wire.

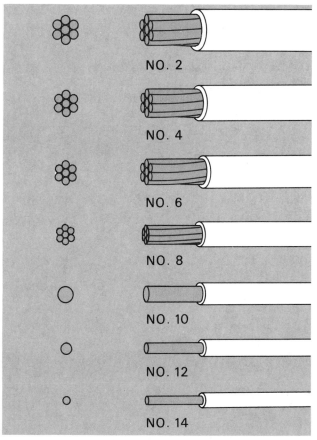

NO. 2

NO. 4

NO. 6

NO. 8

NO. 10

NO. 12

NO. 14

FIGURE 134. Available sizes of conductors in nonmetallic sheathed cable are 2 through 14 if copper and 2 through 12 if aluminum or copperclad aluminum.

(a) Type NM nonmetallic-sheathed cable is most often used for house wiring. It may be installed for both exposed and concealed work in normally dry locations. This includes air voids in masonry block or tile walls if excessive moisture is not present. Type NM may **not** be installed where it is exposed to corrosive fumes or vapors. Neither can it be imbedded in masonry, concrete or plaster.

(b) Type NMC nonmetallic-sheathed cable, in addition to meeting the requirements for NM, must also be fungus resistant and corrosion resistant. It may be used in both exposed and concealed work in dry, moist, damp, or corrosive locations, and in outside and inside walls of masonry and block or tile. Type NMC is not readily available; however, Type UF Cable is available and will meet the requirements of NMC Cable installations.

(5) CABLE IDENTIFICATION. Every cable must be marked to show the following (Figure 135):

—Letters indicating the type of wire.
—Maximum working voltage.
—Manufacturer's name, trademark or other identifying marking.
—The AWG or circular mil size. Surface markings shall appear at least every 61 cm (24 inches).

(6) CABLE AMPACITY. Table VI gives allowable ampacities of different types of cable, with not more than three insulated conductors in a cable or conduit, AWG sizes 18–2000 KCMiL. The figures apply to conductors in a cable or to those directly buried in the earth.

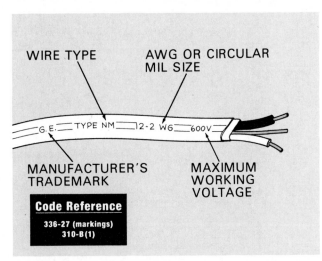

FIGURE 135. Cable identification.

2. REMOVING CABLE FROM THE ROLL

Nonmetallic-sheathed cable is brought to the job as a roll or coil inside a box or on a reel. The way it is removed from the roll can make a difference in the ease of handling.

Pulling a length of cable for a long run is usually easier if you unwind the cable from the inside of the roll. Most cartons have a circle marked to be cut out for the opening. Proceed as follows:

1. *Rotate the end of the cable against the direction of the coil as you remove it from the inside of the roll (Figure 136).*

2. *Cut the length required to reach from the SEP to the first box in the circuit run.*

   If using a reel, pull from outside of reel to the outlet and cut needed length.

   Cable may be spliced only inside a box. Be sure you allow enough wire to reach from the SEP to the first box and from one box to another in each circuit. Also, at each box allow 15 to 20 cm (6 to 8 inches) of cable for making splices and connections inside the box. Much longer lengths are needed for connections inside the SEP cabinet, 61 to 91 cm (24 to 36 inches).

   NOTE: No device connections will be made at either the SEP or the boxes until all plastering, wallboard or paneling is in place. Damage to switches and receptacles is less likely if installed after plaster and/or wallboard are in place.

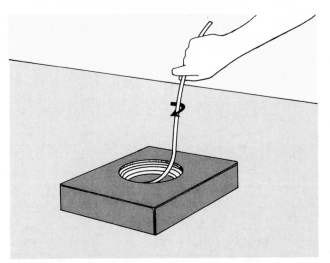

FIGURE 136. Rotate the cable against the direction of the coil as you remove it from inside the coil.

TABLE VI. ALLOWABLE AMPACITIES OF INSULATED CONDUCTORS
RATED 0–2000 VOLTS
(From Code Table 310-16)

Allowable Ampacities of Insulated Conductors
Rated 0–2000 Volts, 60° to 90°C (140° to 194°F)
Not More Than Three Conductors in Raceway or Cable or Earth
(Directly Buried), Based on Ambient Air Temperature of 30°C (86°F).

Size	Temperature Rating of Conductor. See Table 310–13.						Size
	60°C (140°F)	75°C (167°F)	90°C (194°F)	60°C (140°F)	75°C (167°F)	90°C (194°F)	
AWG kcmil	TYPES †TW, †UF	TYPES †FEPW, †RH, †RHW, †THHW, †THW, †XHHW, †THWN, †USE, †ZW	TYPES TA, TBS, SA, SIS, †FEP, †FEPB, MI, †RHH, RHW-2, THW-2, †THHN, †THHW, THWN-2, USE-2, XHH, †XHHW, XHHW-2, ZW-2	TYPES †TW †UF	TYPES †RH, †RHW, †THHW, †THW, †THWN, †XHHW, †USE	TYPES TA, TBS, SA, SIS, †THHN, †THHW, THW-2, THWN-2, †RHH, RHW-2, USE-2, XHH, XHHW, XHHW-2, ZW-2	AWG kcmil
	COPPER			ALUMINUM OR COPPER-CLAD ALUMINUM			
18	. . . .	. . . .	14	. . . .	. . . .	. . . .	. . . .
16	. . . .	. . . .	18	. . . .	. . . .	. . . .	. . . .
14	20†	20†	25†	. . . .	. . . .	. . . .	. . . .
12	25†	25†	30†	20†	20†	25†	12
10	30	35†	40†	25	30†	35†	10
8	40	50	55	30	40	45	8
6	55	65	75	40	50	60	6
4	70	85	95	55	65	75	4
3	85	100	110	65	75	85	3
2	95	115	130	75	90	100	2
1	110	130	150	85	100	115	1
1/0	125	150	170	100	120	135	1/0
2/0	145	175	195	115	135	150	2/0
3/0	165	200	225	130	155	175	3/0
4/0	195	230	260	150	180	205	4/0
250	215	255	290	170	205	230	250
300	240	285	320	190	230	255	300
350	260	310	350	210	250	280	350
400	280	335	380	225	270	305	400
500	320	380	430	260	310	350	500
600	355	420	475	285	340	385	600
700	385	460	520	310	375	420	700
750	400	475	535	320	385	435	750
800	410	490	555	330	395	450	800
900	435	520	585	355	425	480	900
1000	455	545	615	375	445	500	1000
1250	495	590	665	405	485	545	1250
1500	520	625	705	435	520	585	1500
1750	545	650	735	455	545	615	1750
2000	560	665	750	470	560	630	2000

CORRECTION FACTORS

Ambient Temp.°C	For ambient temperatures other than 30°C (86°F), multiply the ampacities shown above by the appropriate factor shown below						Ambient Temp. °F
21-25	1.08	1.05	1.04	1.08	1.05	1.04	70-77
26-30	1.00	1.00	1.00	1.00	1.00	1.00	78-86
31-35	.91	.94	.96	.91	.94	.96	87-95
36-40	.82	.88	.91	.82	.88	.91	96-104
41-45	.71	.82	.87	.71	.82	.87	105-113
46-50	.58	.75	.82	.58	.75	.82	114-122
51-55	.41	.67	.76	.41	.67	.76	123-131
56-60	. . . .	.58	.71	. . . .	.58	.71	132-140
61-70	. . . .	.33	.58	. . . .	.33	.58	141-158
71-80	. . . .	. . . .	.41	. . . .	. . . .	.41	159-176

†Unless otherwise specifically permitted elsewhere in this Code, the overcurrent protection for conductor types marked with an obelisk (†) shall not exceed 15 amperes for 14, 20 amperes for 12, and 30 amperes for 10 copper; or 15 amperes for 12 and 25 amperes for 10 aluminum and copper-clad aluminum after any correction factors for ambient temperature and number of conductors have been applied.

3. PULLING THE CABLE

When pulling cable, be careful not to damage insulation. This is where holes drilled in line, or with gradual changes in elevation make the job easier. When turning corners, the loop must not be pulled too tight. The insulation will break if turns are too short (Figure 137). The Code states that no bend shall have a radius less than 5 times the diameter of the cable.

Pulling cable is discussed under the following heading.

   a. General Procedures for Pulling Cable.

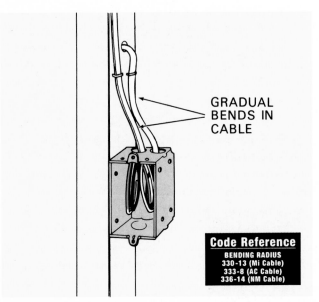

GRADUAL BENDS IN CABLE

Code Reference
BENDING RADIUS
330-13 (Mi Cable)
333-8 (AC Cable)
336-14 (NM Cable)

FIGURE 137. Make gradual bends in cable to prevent damage to insulation.

a. General Procedures for Pulling Cable

Proceed as follows:
1. *You may start at the SEP or at a box to be wired.*
   In this example, assume you will start at the SEP pulling No. 12-2w/g.

   Allow enough wire at the SEP to go around the inside of the box to reach the terminals when installed (Figure 138).

2. *Pull the cable from the coil through holes, notches, alongside guard strips or over running boards.*

   Pull the cable through each opening to the outlet or switch box.

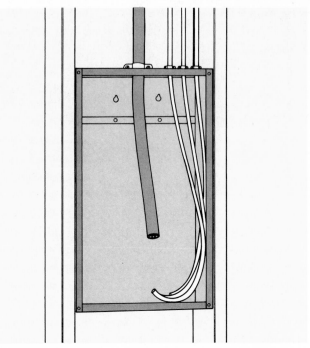

FIGURE 138. Allow enough cable for connecting inside the SEP.

3. *Pull the cable through the knockout hole in the box.*

   Allow an extra 15 to 20 cm (6 to 8 inches) for splicing and connections.

4. *Cut cable for correct length at the roll.*

5. *Remove cable sheathing.*

   Many prefer to remove the **cable sheathing** at this time and fold the individual wires back into the box. Insulation will be removed from the ends of wire later. Figure 139 shows sheathing removed from cables in the box. To remove the sheathing leave 0.6 to 2.5 cm (¼ to 1 inch) of the cable sheathing exposed past the clamp on the inside of the box (Figure 139).

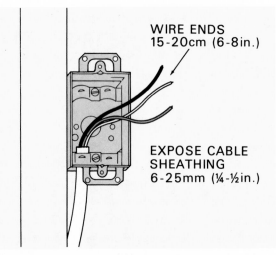

WIRE ENDS
15-20cm (6-8 in.)

EXPOSE CABLE SHEATHING
6-25mm (¼-½ in.)

FIGURE 139. Leave .6 to 2.5 cm (¼ to 1 inch) of cable sheathing exposed past the clamp inside the box.

The sheathing can be cut with a knife. Some prefer to use a cable ripper. Instructions for use are given in the following section, "Connecting Circuits."

6. *Fasten cable in boxes.*

Cable must be securely fastened to every **metal** box. The cable is not required to be fastened to a **nonmetallic** box no larger than 5.7 × 4 cm (2¼ × 4 inches) when it is secured with a strap or staple within 20 cm (8 inches) of the box. For larger nonmetallic boxes, the cable must be fastened to the box. For requirements for securing cable in nonmetallic boxes, check Code Section 370-17 (c) Ex. and Code Section 370-17 (b) for metallic box applications.

Two types of anchors (fasteners) are used for metal boxes. They are:

(a) Connectors.

(b) Clamps.

(a) To install cable using connectors, proceed as follows:

(1) *Slide the connector over the cable 15 to 20 cm (6 to 8 inches) from the end and tighten locking screws (Figure 140).*

This allows 15 to 20 cm (6 to 8 inches) of cable inside the box for connections.

(2) *Insert cable and connector through the knockout and install locknut (Figure 140).*

(3) *Tighten locknut with screwdriver (Figure 141).*

Some prefer to install the connector through the knockout first, then install cable and tighten connector screws.

(4) Several UL-listed **plastic** connectors are now marketed that snap into a knockout. For example, one type of plastic connector does not require any tools for installation. It snaps into the knockout (Figure 142). When the cable is inserted, the connector clamps it tightly. Another type snaps into place and a

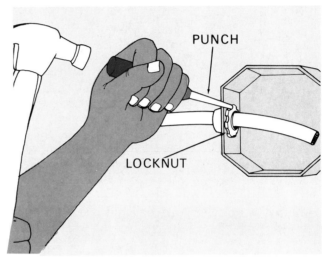

FIGURE 141. Tightening locknut.

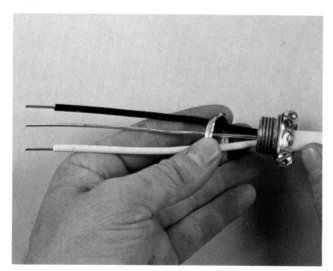

FIGURE 140. Cable and connector to be installed through knockout.

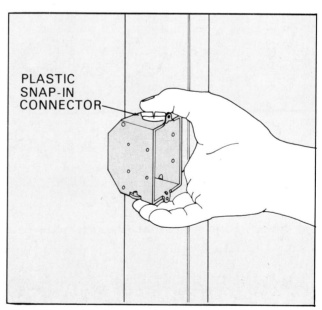

FIGURE 142. Plastic snap-in connector.

small locking wedge fastens the cable when squeezed with pliers (Figure 143). A third type snaps in place and locks the cable with the turn of a screwdriver (Figure 144). Still another snap-in type uses the screwdriver as a lever to lock a wedge against the cable (Figure 145).

(b) **To install cable using clamps, proceed as follows:**

(1) *Insert cable under clamp leaving 15 to 20 cm (6 to 8 inches) beyond the clamp for connections and tighten bolt (Figure 146).*

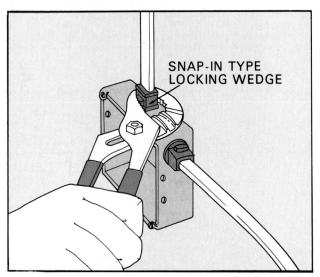

FIGURE 143. Small locking wedge fastens cable when clamped firmly with pliers.

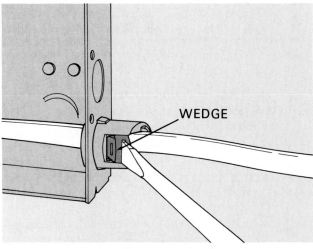

FIGURE 145. Screwdriver used as a lever to lock the wedge against the cable.

FIGURE 144. A turn of the screwdriver snaps wedge into place.

FIGURE 146. Pull 15 to 20 cm (6 to 8 inches) of cable beyond the clamp and tighten the bolt or screw.

(2) If "quick-clamp" type, pry up edge of clamp and pull cable through (Figure 147).

7. Pull cable for remaining boxes in the circuit as follows (Figure 148):

(1) Move roll or pull cable end to first box.

(2) Pull cable through holes, notches or guard strips to second box.

Pull cable through knockout and fasten.

(3) Cut cable at the roll.

(4) Insert cable end through a knockout in first box and fasten with connector or clamp.

(5) Repeat for each box until circuit is complete.

NOTE: The bathroom receptacle in Figure 148 is installed in a circuit with the receptacle in the adjacent bathroom and may be connected to a GFCI in the SEP or individual GFCI's may be installed.

8. Pull cable to switch boxes (Figure 149).

The Code calls wiring from light to switch box the **switch loop**. Two-wire cable **with ground** should be used for most switch loops. Run the cable to each switch connection from the ceiling light outlet box through the plate to the switch box.

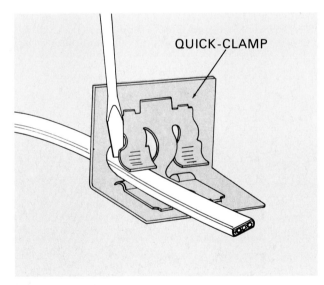

FIGURE 147. Quick-clamp type cable clamp.

NOTE: Some switch installations require the use of a cable with more than two conductors. Switch installations are discussed under Section III-E, "Connecting Switches and Circuits."

9. Identify the branch circuit at the SEP.

Service entrance panel covers usually have a numbered chart inside the door on which each circuit can be identified. Label each circuit as you pull the cable to simplify the completion of the chart later.

Attach masking tape to cable end in the SEP and label the circuit. For example, Circuit No. 2, Bedroom No. 1, and Bathroom.

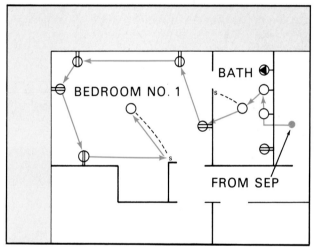

FIGURE 148. Pull cable to lighting outlets, receptacles and switches indicated for circuit No. 2.

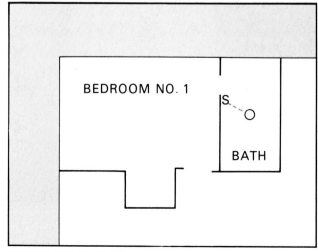

FIGURE 149. Switch loop for bath.

G. Anchoring Cable

Cable must be anchored or fastened as specified by the Code on surfaces over which it runs. Anchoring cable is discussed as follows:

1. Types of Anchors.
2. Strapping or Stapling Cable.

1. TYPES OF ANCHORS

The Code requires cable to be anchored by staples, straps or similar fittings so designed and installed as not to injure the cable (Figure 150). The use of staples is permitted but they must be installed carefully. The cable sheathing and wire insulation will likely be damaged if a staple is bent or driven too far.

2. STRAPPING OR STAPLING CABLE

Where cable is run through holes drilled in framing members, you are not required to fasten nonmetallic sheathed cable except within 30 cm (12 inches) of the box (Figure 151).

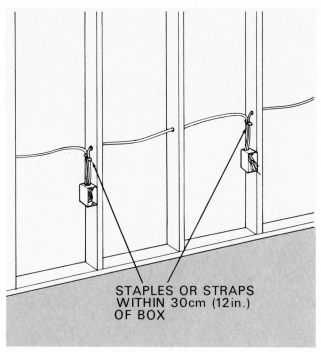

FIGURE 151. Nonmetallic sheathed cable must be fastened within 30 cm (12 inches) of every metal box.

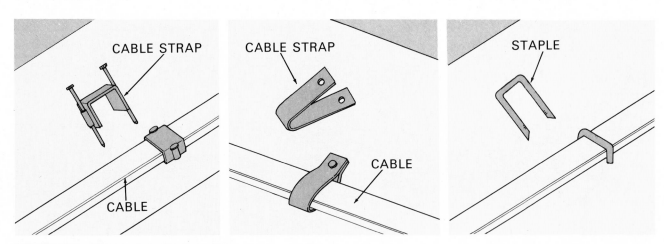

FIGURE 150. Means of anchoring cable.

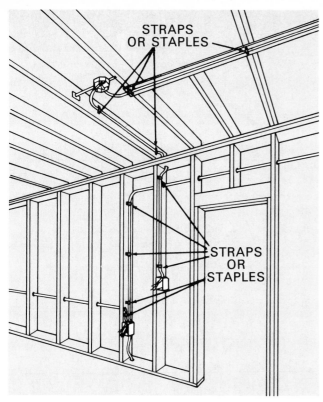

All cable on studs or run across the face of framing members and at right angles to joists, rafters and sills must be fastened at least every 1.4 m (4½ feet) and within 30 cm (12 inches) of every metal box (Figure 152). Cable must be fastened **within 20 cm (8 inches)** of nonmetallic boxes if cable connectors or clamps are not used at the box. If multi-gang (more than one) nonmetallic boxes are installed, you must fasten the cable at the box.

Where notches in framing are covered with steel plates, the cable is considered secured. But you must still fasten the cable within 30 cm (12 inches) of every metal box and within 20 cm (8 inches) of nonmetallic boxes (Figure 153).

FIGURE 152. Cable fastened to studs run across framing members and at right angles to framing.

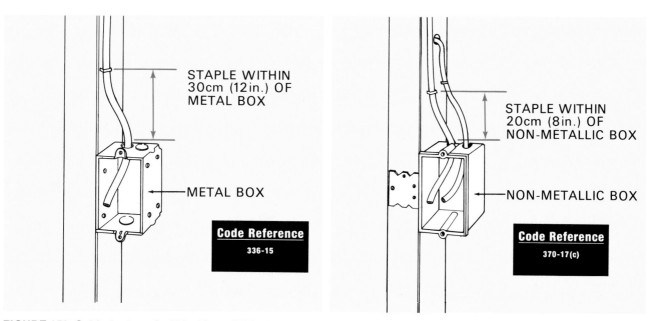

FIGURE 153. Cable fastened within 30 cm (12 inches) of every metal box and within 20 cm (8 inches) of nonmetallic boxes.

NOTES

III. Connecting Circuits

The service entrance panel box or cabinet should have the cable ends neatly in place around the sides of the box after cable has been pulled to the outlets (Figure 154). The open space in the panel box will be occupied by the service entrance wires, the neutral bar, and the panel for the snap-in circuit breakers or fuses (Figure 155). After all wallboard and plaster are in place, your next task is to connect the wiring to the device in each box.

A new section in the 1993 Code (Section 110-12 [c]) is designed to protect electrical equipment during construction. Often, the internal parts of electrical equipment are damaged by craftsmen who do not realize the long-range consequences of their actions. Section 110-12 (c) provides the authority having jurisdiction over enforceable requirements information to protect **internal parts of electrical equipment**.

Upon successful completion of this section, you will be able to **use correct procedures to connect all circuits required for wiring a house, including receptacles, switches, lighting outlets, and the required grounding**. Connecting circuits is discussed under the following headings:

A. Connecting Conductors.
B. Grounding the Electrical System and Equipment.
C. Connecting Circuits and Receptacles.
D. Installing Lighting Fixtures.
E. Connecting Switches and Circuits.
F. Installing Doorbells, Chimes, and Smoke Detectors.

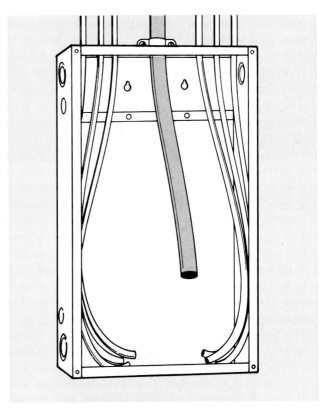

FIGURE 154. Ends of cable for each circuit pulled into the SEP box (cabinet).

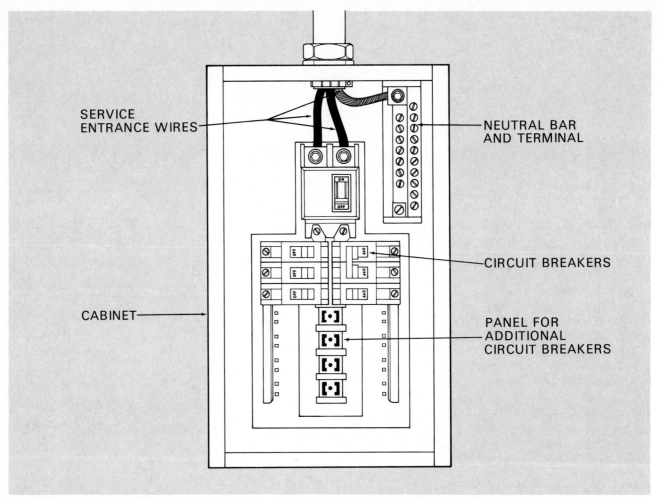

FIGURE 155. SEP box with service entrance wires, neutral bar and circuit breaker panel installed.

Labels in figure:
SERVICE ENTRANCE WIRES
NEUTRAL BAR AND TERMINAL
CIRCUIT BREAKERS
CABINET
PANEL FOR ADDITIONAL CIRCUIT BREAKERS

A. Connecting Conductors

A dependable wiring job depends to a great extent on good connections throughout the system. Wherever wires are attached to terminals or spliced together, take time to make the connection correctly and prevent problems in the future.

Connecting conductors is discussed as follows:
1. Preparing Cable for Connections.
2. Types of Splices and Connectors.
3. Splicing Wires.

1. PREPARING CABLE FOR CONNECTIONS

The first step in preparing cable for the connection in boxes is to remove the outer covering called a sheath, if not removed before installing in the box. A cut down the center of the cable will open the sheath making it easy to separate from the insulated wires inside. A knife blade will do the job but may cut through the insulation on the wires. Many prefer to use a cable ripper because it does not usually penetrate deep enough to cut the wire insulation (Figure 156).

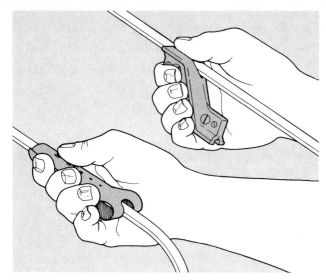

FIGURE 156. Cable ripper used to open the sheath on nonmetallic sheathed cable.

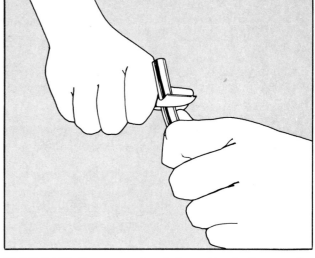

FIGURE 157. To remove insulation with a knife, start the cut at angle to prevent a nick in the wire.

Preparing cable for connections is discussed under the following headings:

a. Removing Cable Sheath.
b. Removing Insulation from Wires.

a. Removing Cable Sheath

Proceed as follows to remove cable sheath with cable ripper:

1. *Place the cutting edge of the tool where you want the cut to start on the cable.*
2. *Squeeze and pull to the end of the cable.*
3. *Peel the sheathing and cut it loose with knife or side cutter.*
4. *Remove any remaining paper and other filler.*

b. Removing Insulation from Wires

Insulation may be removed with a knife or a wire stripping tool. Proceed as follows:

1. *If you use a knife, start the cut on the insulation at an angle to prevent cutting into the wire (Figure 157).*

   A nick in the wire will weaken it and reduce current carrying capacity. Remove about 1.5 to 2 cm (⅝ to ¾ inch), depending on type of connection to be made.

2. *If you use a wire stripper, insert the wire into the correct wire gage hole size on the tool and squeeze the handles.*

   Be sure the hole size is correct or you will damage the wire. Squeeze to make the cut through the insulation and strip the insulation off the end of the wire (Figure 158). This method does not cut into the wire.

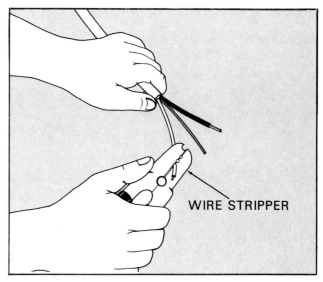

WIRE STRIPPER

FIGURE 158. To remove insulation with a wire stripper, match stripper notch to wire size, insert the wire, squeeze the handle and strip the insulation off the wire.

2. TYPES OF SPLICES AND CONNECTORS

Splicing is the joining together of two or more wires to make a connection. Splices must conduct electricity as well as an uncut wire to be sure of good electrical connections. Loose connections may allow arcing which could cause a fire or damage equipment when the system is placed in service.

Soldering was once the accepted method of splicing wires but is seldom used in modern house wiring. Soldering is **not** permitted for connecting the grounding electrode conductor to the grounding electrode.

Solderless connectors are available for most types and sizes of splices (Figure 159). They are easy to use and make good connections quickly. There are several brands but they are usually called **wire nuts,** which is the name used in this book (Figure 160). Solderless

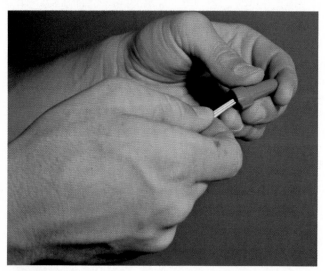

FIGURE 159. Solderless connector for quick and easy splicing.

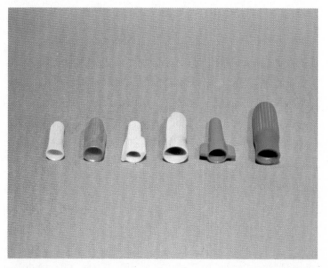

FIGURE 160. Wire nuts of various types, size and color.

connectors are used for practically all splices in house wiring circuits. They are available in sizes to fit all house wiring. The size of wire must be matched with correct wire nut size given on the package. Another type of solderless connector is the crimp-type made of plastic covered metal. **Split-bolt** connectors of various types and sizes are used to splice heavier wires (Figure 161).

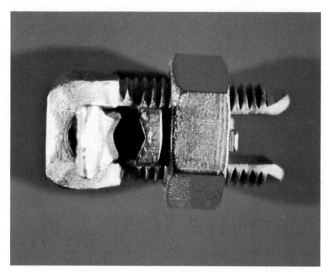

FIGURE 161. Split-bolt for splicing heavy conductors.

3. SPLICING WIRES

Remember, all splices must be enclosed in a box. There are two methods for splicing wires:
   a. Solderless.
   b. Soldered.

a. Solderless

Solderless splices may be made by the use of **three** different types of connectors. They are described as follows:

(1) WIRE NUT CONNECTORS are the most satisfactory. Proceed as follows:
   1. *Remove insulation from wire ends.*
      Use a wire stripper.
      Remove just enough insulation for the connector to cover the bare wire when tight (Figure 162). Follow instructions on the box.

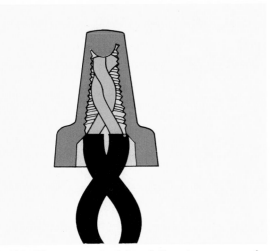

FIGURE 162. The amount of insulation to remove depends on size of wire and wire nut.

2. *Hold wires side by side.*

It is not necessary to twist the wires together.

3. *Place connector over wire end and turn as you would a nut until tight (Figure 163).*

Be sure both wires are locked in and no bare wires are exposed.

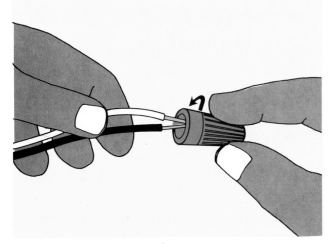

FIGURE 163. Attaching a wire nut (solderless connector).

(2) CRIMPED-CONNECTORS are used but may not be approved by some local codes. The wire ends are stripped, placed in a connector and crimped with a crimping tool (Figure 164).

(3) SPLIT-BOLT CONNECTORS are normally used for large wires (Figure 165). The wire ends are stripped, placed in the connector and the nut is tightened. The connection is then wrapped with layers of plastic electrical tape at least the same thickness as the original insulation.

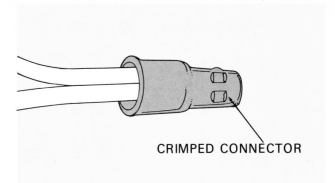

FIGURE 164. A crimp-connector splice.

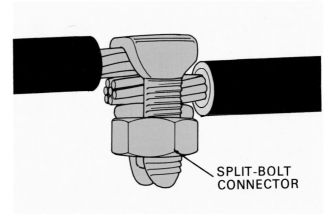

FIGURE 165. A split-bolt connector splice. Wrap with tape to same thickness as wire insulation.

b. Soldered

Wires may be spliced with solder. Wire ends are stripped, cleaned and twisted together. The wire ends are heated with a soldering iron and rosin-core solder is melted by the heated wire (Figure 166).

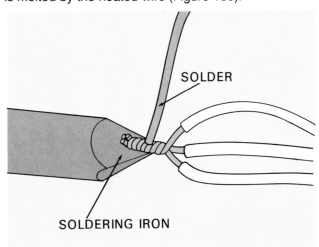

FIGURE 166. Soldering a splice. Heat wires with soldering iron underneath to melt solder held on top.

B. Grounding the Electrical System and Equipment

A **ground** is an intentional connection between an electrical system and the earth or some other conducting body that serves in place of the earth. The preferred way to ground a house wiring system formerly was to connect a conductor from the neutral bar in the SEP to a metal cold water pipe in direct contact with the earth for at least 3.1 m (10 feet) as indicated by Code Section 250-81 (a) (Figure 167). This type of installation must now be supplemented by one or more additional grounding electrodes. Details on conductor sizes and other methods of grounding are given under Section IV, "Installing Service Entrance Equipment." Grounding electrode conductor connection to the metallic water pipe must be made within 5' of the point of the metal pipe entering the structure, according to Code Section 250-81.

Grounding makes an electrical system safer for people to use and prevents damage to the system and property. If circuits and equipment are not correctly grounded, you may get a severe or even fatal shock. Such items as appliances, power tools and electric motors may be damaged or destroyed by a ground fault. A ground fault is an insulation failure between an energized (hot) conductor and ground.

Grounding the electrical system and equipment is discussed as follows:

1. Grounding the System and Circuits.
2. Grounding the Equipment and Conductor Enclosures.

1. GROUNDING THE SYSTEM AND CIRCUITS

Grounding the system and circuits covers all current-conducting parts of the electrical installation, including the necessary wiring in the SEP and the circuits. In this discussion, you are concerned primarily with grounding the house wiring circuits from the SEP to all parts of the house.

System and circuit grounding is accomplished by connecting the SEP neutral bar to ground as discussed earlier (Figure 167) and by connecting the white neutral wire from each circuit to the neutral bar (Figure 168). In some localities two bars are required. One bar is used to connect the white neutral wire in the cable and one for the grounding wire. Both bars are connected to ground through a screw or bolt (if metal); otherwise, is must be connected to the grounding conductor feeding into the SEP.

In 120-volt installations there must be two conductors, one black and one white. The black wire is commonly referred to as the "**hot**" wire, but its technical name is the **ungrounded conductor**. The white wire is commonly called the **neutral wire**, but its technical name is **grounded circuit conductor**. Both wires carry current, but the white neutral wire is connected to ground through the neutral bar in the SEP (Figure 168). The white neutral wire must be continuous throughout the system. How can this be done when the wire must be cut at every box or outlet to make spliced connections?

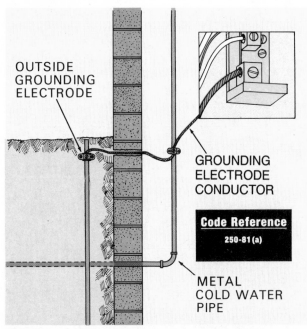

OUTSIDE GROUNDING ELECTRODE

GROUNDING ELECTRODE CONDUCTOR

Code Reference
250-81(a)

METAL COLD WATER PIPE

FIGURE 167. The metal water-pipe grounding electrode supplemented by additional electrode to conform to the Code. Grounding electrode connection must be made within 5 ' of the point of entrance of metallic pipe into the structure, according to Section 250-81 of the Code.

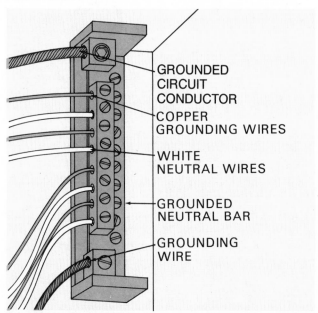

GROUNDED CIRCUIT CONDUCTOR

COPPER GROUNDING WIRES

WHITE NEUTRAL WIRES

GROUNDED NEUTRAL BAR

GROUNDING WIRE

FIGURE 168. White neutral wires must be connected to grounded neutral bar in the service entrance panel for each circuit.

It can be made electrically continuous by joining the cut ends with good solid connections at every box (Figure 169). By joining white wire to white wire at every connecting point, the effect is the same as having one continuous white wire. It runs from the SEP neutral bus to every outlet in the circuit. **You must not install a switch, a fuse, or a circuit breaker in the neutral wire.** The rule to remember is, **the neutral is always white but the white wire is not always a neutral.** The white wire is not a neutral if it is connected to a switch. In an exception to the code requirements, the Code allows a white wire to be connected to a switch in some instances (see Code Section 200-7). The exception is explained in detail in Section E, "Connecting Switches and Circuits."

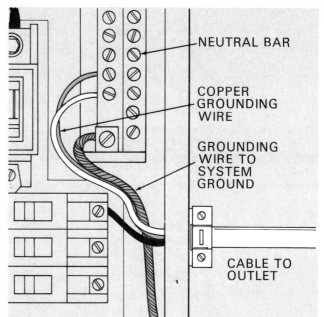

FIGURE 170. The grounding wire must be connected to the neutral bar in the SEP.

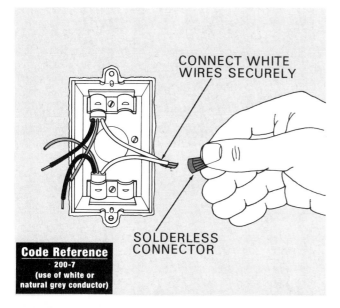

FIGURE 169. The white neutral wire must be made continuous by spliced connections at each box or outlet in the circuit.

2. GROUNDING THE EQUIPMENT AND CONDUCTOR ENCLOSURES

Equipment grounding covers the non-current-carrying parts of the system. Included are metal frames on equipment, appliances, and motors, and all metal outlet boxes, switch boxes, receptacles and metal conduit.

According to the Code, equipment is connected to ground for the following purposes:

— To **limit the voltage to ground** on equipment frames and on metal enclosures of the wiring system.
— To **assure operation of over-current devices** in case of ground faults.

The grounding wire is there to help protect a person from a dangerous shock or to protect equipment from damage in case of insulation failure in the circuit. If a ground fault occurs and the current in the line exceeds its overcurrent protection, the circuit breaker trips or the fuse opens, disconnecting the circuit.

In 120-volt house wiring using nonmetallic sheathed cable w/ ground, **equipment grounding** is accomplished by means of the bare grounding wire in the cable. It must be connected to the neutral bar (Figure 170). The bare grounding wire does not normally carry current. From the neutral bar, it must be connected to metal outlet boxes, to every outlet in the circuit and to the frames of all equipment. The grounding wire may sometimes be insulated. If so, it must be green or green with yellow stripes, according to Section 210-5 (b) of the Code.

By connecting the white neutral wire and the bare grounding wire to the neutral bar, you have grounded both the circuit and the equipment. The following example illustrates grounding the clothes washer and its circuit (Figure 171). Note the terminal connections on the back of the washer. The black and white wires of the cord carry current to operate the washer. When the cord is plugged into the receptacle, the neutral wire is connected to ground on the neutral bar through the white wire in the circuit cable. An accidental contact of the non-grounded current-carrying wires to circuit ground will result in a short circuit, and the circuit breaker or fuse will open to break the circuit.

The green grounding wire in **Figure 171** is fastened to the **frame** of the washer by means of a metal screw terminal. When plugged into the receptacle, the grounding wire is connected by the grounding circuit to the neutral bar. This means that the **equipment** is **grounded** through the grounding wire to the neutral bar. Thus, if a non-grounded circuit conductor should make contact with the frame or machinery of the washer, a short circuit fault will occur and open the fuse or trip the circuit breaker. Grounding the outlet boxes and receptacles has the effect of grounding all applicances and equipment that may be plugged into any receptacle on each circuit, provided the appliance cord has a grounding conductor and plug.

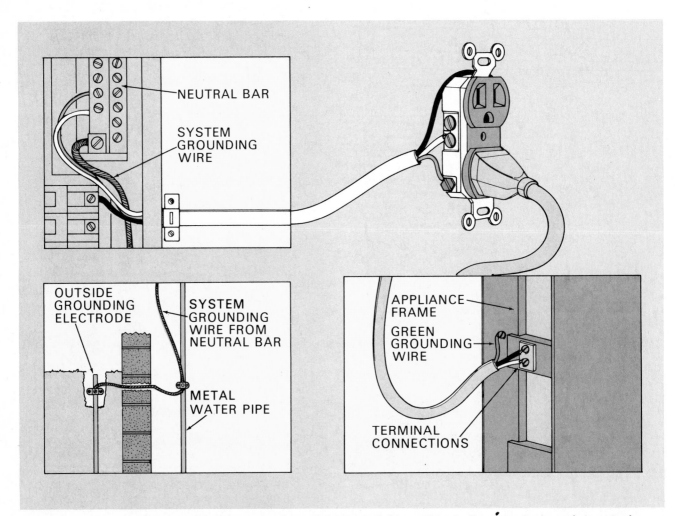

FIGURE 171. The grounding wire connects to the metal frame of the washer, to the receptacle and receptacle box and to the netural bar in the service entrance panel.

To ground the system and circuits and the equipment, proceed as follows (Figure 172):

1. *In the SEP, locate the end of the cable marked "clothes washer."*

2. *Remove the sheath and filler from cable clamp to the ends of the cable.*

3. *Remove 2 cm (¾ inch) of insulation from each insulated wire.*

4. *Insert the white neutral wire under a terminal screw on the neutral bar and tighten. On most neutral bars the wire-end is inserted under the terminal screws without a loop.*

5. *Insert the bare grounding wire under a separate terminal screw on the neutral bar and tighten the screw.*

To complete the circuit wiring in the SEP, connect the black wire to the circuit breaker. Most have push-in set-screw type connections. Snap the circuit breaker into position. Be sure it is the correct size for the conductor.

6. *Connect the grounding wire to the metal box and to the receptacle grounding screw.*

The bare grounding wire in the cable must be connected to the metal outlet box and to the green grounding screw on the receptacle (Figure 173A). Code Section 250-74 provides information on the usual way to connect the grounding wire to a metal box through the use of a "jumper" wire, sometimes called a "pigtail," and a screw (Figure 173A). The "jumper" is a short piece of wire, 10–15 cm (4–6 inches) long. Another method of grounding the box is by means of a clip. Insert the end of the jumper wire into the clip and force the clip onto the edge of the box (Figure 173B). Use of clips is not allowed by some local codes.

In single-gang plastic boxes, the grounding wire is not required to be fastened to the box. The grounding wires are fastened to each other and to the grounding terminal on the receptacle (Figure 174).

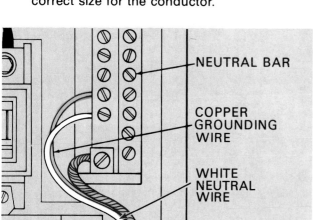

FIGURE 172. Connect the black wire in the cable to the circuit breaker.

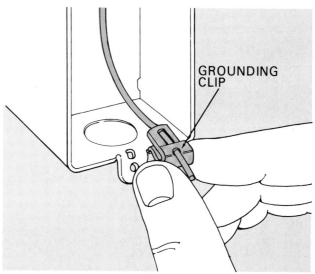

FIGURE 173B. A grounding clip fastens the jumper wire to the edge of the box.

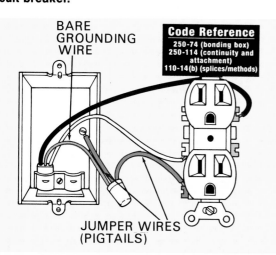

FIGURE 173A. Connect the bare grounding wire from the cable to the receptacle grounding screw and to the box, using two jumper wires and a wire nut solderless connector.

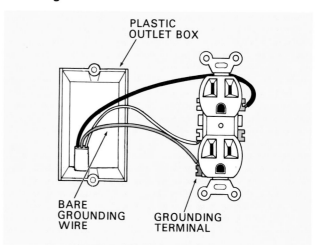

FIGURE 174. For single-gang plastic box, a grounding wire to the box is not required.

The Code permits the following exceptions to the required jumper wire for receptacle grounding:

(a) **Surface mounted boxes.**

Where the box is surface mounted, the metal to metal contact between the metal receptacle yoke and the box is permitted for grounding the box to the yoke (Figure 175).

(b) **Flush-mounted boxes.**

Contact receptacles or yokes designed and **listed** for the purpose are permitted for grounding the circuit between the yoke and flush-type boxes (Figure 176). Several manufacturers offer specially constructed receptacles which provide a ground between the receptacle yoke and box, thus eliminating the grounding screw and jumper wire (Figure 176).

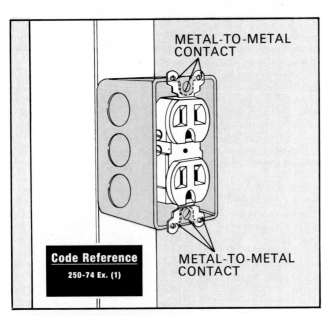

FIGURE 175. Surface-mounted boxes having metal-to-metal contact between the receptacle and box are considered to be grounded.

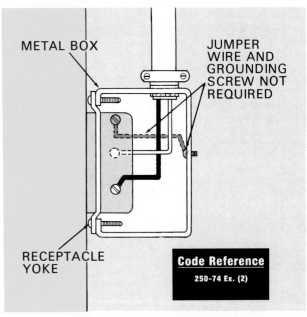

FIGURE 176. Receptacles with yokes that are designed and listed for the purpose may be installed without a grounding wire to the box.

C. Connecting Circuits and Receptacles

In this section, you will connect circuits for each type of receptacle. Connecting circuits and receptacles is discussed as follows:

1. Types, Sizes and Colors of Receptacles.
2. Connecting Receptacles.

1. TYPES, SIZES AND COLORS OF RECEPTACLES

The most common type of receptacle in house wiring is the standard grounding type duplex receptacle. It is available in 15- and 20-ampere, 120-volt rating with the U-shaped opening as the grounding terminal (Figure 177).

The same receptacle is available with a decorator face (Figure 178) and with cover plates in decorator colors including white, brown, grey, ivory, pink, blue, yellow and black. You may also purchase receptacles designed for 2-circuit installation. Another type allows one outlet to be wired for 15-ampere, 125 volts and one for 15-ampere, 250 volts (Figure 179). Receptacles may be rated for **125** and **250** volts even though they are connected to **120** and **240** volts.

Other types and sizes of receptacles are the single receptacle type, single receptacle with a switch, single receptacle with pilot light, and a duplex receptacle with weatherproof cover (Figure 180).

Self-contained kitchen ranges normally require a heavy duty 50 ampere, 125/250-volt receptacle. The flush-mounted type is most often used in new construction (Figure 181). A similar 250-volt receptacle is used for clothes dryers that operate on 240 volts but with different configurations (arrangement of openings) because of different ampere ratings (Figure 182).

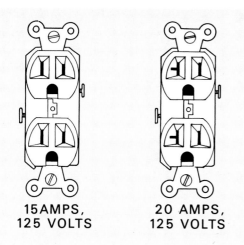

15 AMPS, 125 VOLTS   20 AMPS, 125 VOLTS

FIGURE 177. Standard grounding duplex receptacles.

FIGURE 178. Grounding duplex receptacle with decorator face.

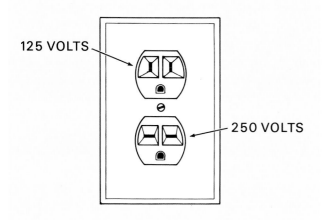

125 VOLTS

250 VOLTS

FIGURE 179. Grounded duplex receptacle rated at 15 amperes, 125 volts in upper outlet and 15 amperes, 250 volts in lower outlet. Note the arrangements of openings, paralleled for 125 volts and tandem for 250 volts.

FIGURE 180. Duplex receptacle with weatherproof cover.

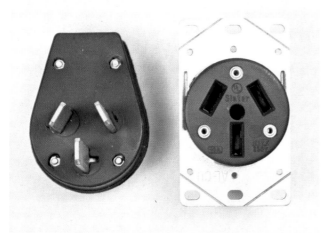

FIGURE 181. Flush-mounted 250-volt, 50-ampere kitchen range receptacle and plug. Also used for some arc welding machines.

FIGURE 182. Dryer receptacle rated at 30 amperes, 250 volts.

Table VII shows the configurations for general purpose, non-locking receptacles and plugs. Each type is also available as a locking plug which provides an extra safety factor. These configurations are approved as the standard by the "National Electrical Manufacturers Association" (NEMA) and are endorsed by Underwriters Laboratories and the electrical industry generally. Note that each configuration is designated by numbers and letters. The range outlet shown in Figure 182 is No. 10-50 R on the chart and the dryer is 10-30 R.

You may also note that most terminals in Table VII are identified by letters. The letter "G" marks the terminal to which an equipment grounding wire is attached. The letter "W" identifies the white neutral wire. If the terminal is unmarked, it is connected to an ungrounded (hot) circuit conductor. On devices containing two or three terminals for hot circuit conductors, the terminals are marked "X," "Y" and "Z."

2. CONNECTING RECEPTACLES

Connecting receptacles is discussed as follows:
  a. Connecting 120-Volt Duplex Receptacles.
  b. Connecting Split-Wired Duplex Receptacles.
  c. Connecting Individual Appliance Circuits.

a. Connecting 120-Volt Duplex Receptacles

Examine the duplex receptacle before you start wiring. Looking at an example from the back (Figure 183), note the two **brass terminal** screws on one side. Always connect a **black** wire or current-carrying wire to a brass terminal (black wire to dark terminal). On the opposite side are the two silver terminal screws. Connect only white wires to silver terminals (white wires to light terminals). The green hexagonal grounding screw is at the bottom on the right. (It may be at the top on some receptacles). Connect the bare or green grounding wire to the green screw.

Some receptacles do not have different colored terminals. If not, the Code requires the word **"white"** to

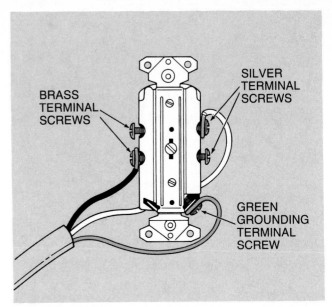

FIGURE 183. Brass-colored terminals for black wires, silver-colored terminals for white wires and green terminal for grounding wire.

be located next to the terminals for the white neutral wires. Also, if the grounding terminal is not colored green, it must have the word **"green"** located next to it.

At the top and bottom center are the mounting screws. They fit the holes on the front of the box. Use these to mount the receptacle after you complete the wiring. Plaster ears at bottom and top hold the receptacle in correct position if the box is set too deep.

Assume you are wiring receptacles in bedroom No. 1 in the standard house plan. No. 12-2 w/g cable is installed and clamped inside the box. You should have about 15 to 20 cm (6 to 8 inches) of cable to work with. Mount the duplex receptacle with U-shaped ground opening at the bottom. Proceed as follows to connect the first receptacle in the circuit:

1. *Remove cable sheath and filler and strip about 1.6 cm (⅝ inch) of insulation from the exposed insulated wires.*

2. *Make a loop of the wire ends of the black and white wires as shown in Figure 125.*

   Leave an opening at the end of each loop so it can be easily placed under the screw terminal.

TABLE VII. NEMA CONFIGURATIONS FOR GENERAL-PURPOSE NONLOCKING PLUGS AND RECEPTACLES

		15 AMPERE		20 AMPERE		30 AMPERE		50 AMPERE		60 AMPERE	
		Receptacle	Plug	Receptacle	Plug	Receptacle	Plug	Receptacle	Plug	Receptacle	Plug
2-POLE 2-WIRE	1 125 V	1-15R	1-15P								
	2 250 V			2-20R	2-20P	2-30R	2-30P				
2-POLE 3-WIRE GROUNDING	5 125 V	5-15R	5-15P	5-20R	5-20P	5-30R	5-30P	5-50R	5-50P		
	6 250 V	6-15R	6-15P	6-20R	6-20P	6-30R	6-30P	6-50R	6-50P		
	7 277 V AC	7-15R	7-15P	7-20R	7-20P	7-30R	7-30P	7-50R	7-50P		
3-POLE 3-WIRE	10 125/250V			10-20R	10-20P	10-30R	10-30P	10-50R	10-50P		
	11 3Φ 250V	11-15R	11-15P	11-20R	11-20P	11-30R	11-30P	11-50R	11-50P		
3-POLE 4-WIRE GROUNDING	14 125/250V	14-15R	14-15P	14-20R	14-20P	14-30R	14-30P	14-50R	14-50P	14-60R	14-60P
	15 3Φ 250V	15-15R	15-15P	15-20R	15-20P	15-30R	15-30P	15-50R	15-50P	15-60R	15-60P
4-POLE 4-WIRE	18 3ΦY 120/208V	18-15R	18-15P	18-20R	18-20P	18-30R	18-30P	18-50R	18-50P	18-60R	18-60P

Configurations approved as NEMA Standard.

81

3. *Connect the black (hot wire) to one of the brass colored terminal screws on the receptacle (Figure 184).*

   Place the loop on the terminal so it will tighten as the screw is turned clockwise. With pliers, pull the loop snug around the screw and tighten with a screwdriver. When tight, the insulation should run right up to the screw so bare wire doesn't show.

4. *Connect the white (neutral) wire to a silver-colored terminal opposite the black wire connection (Figure 184).*

   Install the wire loop under the silver terminal and tighten.

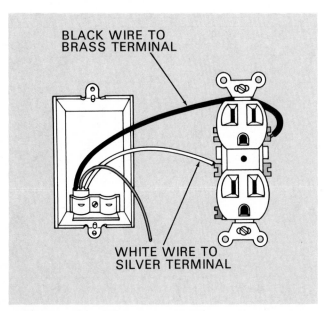

FIGURE 184. Connect black wires to brass terminals and white wires to silver terminals on the receptacle. Place the loop under the screw terminal so it will tighten as the screw is turned.

5. *Connect the grounding wire.*

   You can connect the wires to the receptacle in any order you like. Many people prefer to make the grounding connections first. The grounding wire in the cable is bare or, if insulated, green. The grounding wire must always be connected to every metal box. The grounding wire must be connected to both the box and the receptacle unless the receptacle is the type with special grounding yoke or the box is plastic. Assume you are wiring a receptacle that must be grounded to the box. Two jumper wires connect to the grounding wire(s) in the circuit cable, one from the box and the other from the receptacle.

   Proceed as follows:

   (1) *Connect the jumper wire to the grounding screw in the box.*

      Cut a 15- to 20-cm (4- to 6-inch) length of green insulated or bare wire and connect one end to the grounding screw in the box (Figure 185).

   (2) *Cut a second 15- to 20-cm (4- to 6-inch) length of green insulated or bare grounding wire and connect one end to the green grounding terminal on the receptacle (Figure 185).*

   (3) *Complete the grounding wire connections.*

      Three wire ends must be brought together, the bare wire from the cable and the two jumper wires. Place the three ends of the wires together, attach a wire nut and tighten (Figure 185).

      The receptacle is now completely wired if only one cable enters the box. However, in this instance, you must connect a second outgoing cable to extend the run to other receptacles.

6. *Connect second cable.*

   To connect the second cable, attach the black wire to the second brass terminal screw and

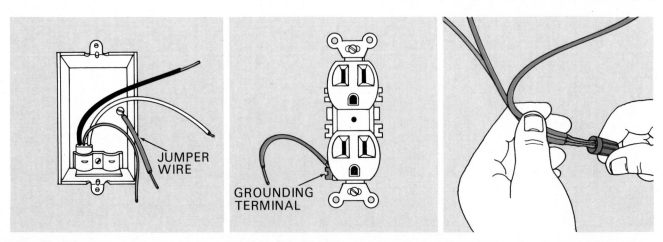

FIGURE 185. Connect one jumper wire to the box grounding screw and one to the receptacle grounding screw. Place the ends of the jumper wires together with the grounding wire from the cable and fasten together with a wire nut.

the white wire to the second silver-colored terminal screw (Figure 186).

Since the grounding connections have already been made to the receptacle and to the box, you need not repeat these steps.

7. *Connect the bare grounding wire from the second cable to the grounding wire from the first cable.*

Instead of three grounding wire ends fastened with a wire nut, you will now have four. Remove the wire nut and place the end of the grounding wire with the ends of the first three. Replace the wire nut and tighten (Figure 186). (On the job you would connect all four grounding wires at once, knowing that the outgoing cable must be connected). This completes connections to the receptacle and box for the incoming cable and the outgoing cable.

8. *Place the receptacle in the box.*

Bend the wires slightly so they will fold in accordion fashion. Push the receptacle into the box with wires folded behind (Figure 187).

9. *Attach the receptacle to the box.*

The receptacle should be mounted so it is straight up and down. The box may sometimes tilt right or left. A wide slot is provided for the mounting screws at top and bottom of the receptacle to permit adjustments as needed (Figure 188).

10. *Attach cover plate.*

After all finish work is completed on wall surfaces, attach the cover plate. Insert center screw to hold cover plate (Figure 189).

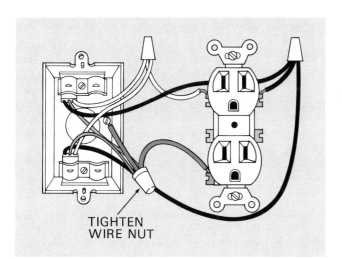

FIGURE 186. Where you have two cables entering the box, place the ends of the two cable grounding wires together with ends of the two jumper wires and fasten with a wire nut.

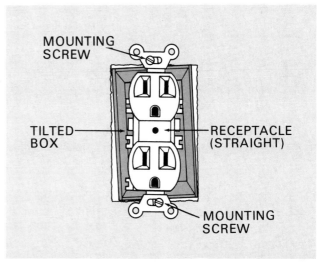

FIGURE 188. Wide slot in strap permits adjustment of receptacle left or right as required for correct position.

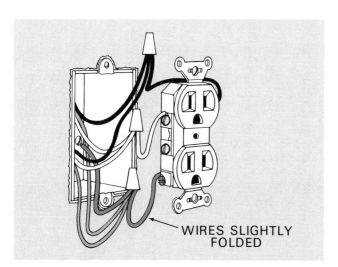

FIGURE 187. Fold wires accordion style to push them into the box more easily.

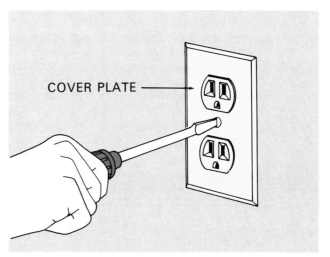

FIGURE 189. Attach cover plate after all plastering and painting is completed.

83

11. *Connect back-wired receptacles.*

Many receptacles are made so wiring can be connected to either side terminals or to back-wiring holes. Connections to back-wired receptacles are similar to side screw terminal connections. The difference is that wires are inserted into holes on the back of the receptacle (Figure 190). Back-wired receptacles have a strip (depth) gage on the receptacle to indicate the depth of the hole. The length of insulation to be removed from the wires must match the length of the strip gage (Figure 190). If you are installing back-wired receptacles, proceed as follows:

(1) *Strip the correct length of insulation from the black and white wires.*

(2). *Insert the wires into the holes (Figure 190).*

With many back-wired devices you need only to insert the wire into the hole near the terminal screw to complete the connection. The wire is clamped when it is inserted. With other back-wired devices, you must first loosen a screw near the hole, insert the wire and tighten the screw against the wire. Follow instructions on the carton or packing slip.

Insert the black wire into the hole near the brass terminal and the white wire into the hole near the silver terminal.

If the screw terminals are not identified by color, the hole for each white neutral wire must have a white ring around it or it must have the word "white" printed next to the hole.

If an outgoing cable for another receptacle is to be connected, repeat the procedure for black and white wires. Insert the wires in the holes provided near the second set of terminals.

CAUTION: Care should be exercised to determine whether back-wired receptacles are suitable to extend conductors. Code Section 300-13 (b) says receptacles (devices) shall not be the means for continuing the circuit. It is recommended that this type of receptacle be used only to terminate a conductor. Any splicing to continue a conductor should be by use of a pigtail and wirenut as shown in the illustrations in this publication. All splices should be in compliance with Code Section 110-14.

A hole may not be provided for the bare grounding wire. Use the green grounding terminal screw if there is not terminal marked green. Make all grounding connections in the same way as with screw terminal receptacles.

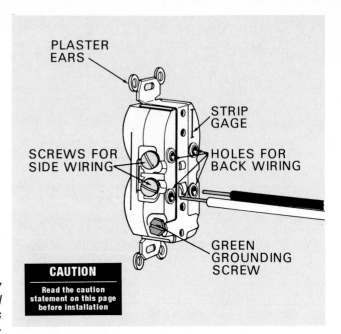

PLASTER EARS

STRIP GAGE

SCREWS FOR SIDE WIRING

HOLES FOR BACK WIRING

GREEN GROUNDING SCREW

CAUTION
Read the caution statement on this page before installation

FIGURE 190. Strip insulation as required and insert wire into holes of back-wired receptacle.

b. Connecting Split-Wired Duplex Receptacles

The standard duplex receptacle has two outlets and two sets of terminals. Both outlets are on the same circuit when wired normally. If you wish to connect a separate circuit to each outlet, you may do so. Most manufacturers attach a **break-off connection tab** between the brass terminals on the receptacle. When the connector tab is removed (Figure 191), the two outlets are no longer electrically connected. That is, they are split. Each outlet can then be connected to a separate circuit. There is seldom any need to remove the tab on the neutral (silver screws) side of the receptacle.

Where do you need a split-wired duplex receptacle? There may be several instances where separate circuits on a duplex receptacle are desirable. One example is in the kitchen appliance circuits. Dividing the load between the upper and lower receptacle outlets prevents possible overloading of a single circuit (Figure 192). In addition, **it is less expensive since two 120-volt circuits are constructed using a 3-wire with ground cable correctly installed to a 120/240-volt circuit.** Only one cable is needed instead of two to obtain two 120-volt circuits. Some workshop circuits may also need split receptacles for shop equipment.

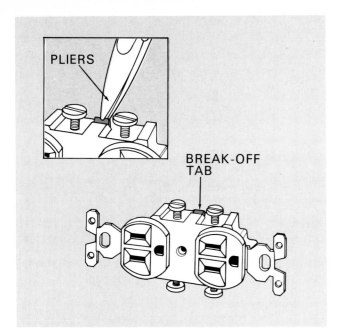

FIGURE 191. Remove the connector tab to wire the duplex receptacle with a separate circuit on upper and lower outlets.

All receptacles wired to this point have been connected to 120-volt, 15- or 20-ampere circuits using 12–2 w/g or 14–2 w/g. To wire the duplex receptacle outlets in the kitchen on separate 120-volt circuits would require two 12–2 w/g cables. In this section, you will follow procedures for wiring split receptacles on separate circuits using one 12–3 w/g cable for a 120/240-volt circuit.

When **two** circuits are wired to the receptacle on a split 240-volt circuit using **one 12–3 w/g cable,** the load must be balanced to avoid problems. To balance

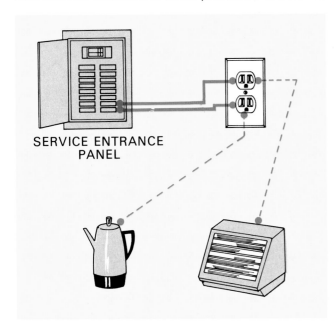

FIGURE 192. Separate (split) circuits for each outlet prevent overloading a duplex receptacle.

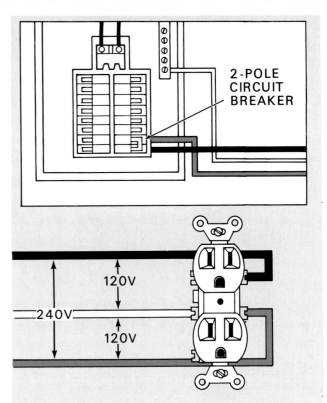

FIGURE 193. A split 240-volt circuit must have a means to disconnect both hot wires at the same time such as a two-pole circuit breaker.

the load, the black wire and red wire must be connected to electrically **opposite legs** of the SEP. On most panels, this is done by placing one above the other (Figure 193).

The Code requires the circuit for a split-wired receptacle to have a double-pole disconnect at the panel board. This may be either a double-pole circuit breaker or two single-pole circuit breakers with a handle-tie (Figure 193). The purpose of this requirement is to be sure both hot conductors are disconnected at the same time. If only one of the two single circuit breakers protecting the 120/240-volt circuit is turned off, the other hot wire could cause severe shock to a person who is working on the receptacle if contact is made. Where the circuit is protected by fuses, install a pullout fuse block to insure that both hot wires are disconnected at once.

When the two hot wires of the circuit are connected to a double-pole circuit breaker or to two single-pole circuit breakers with a handle tie or to a fuse block, the load is divided between two opposite legs of the 120/240-volt supply in the panel board (Figure 193). The load is balanced between the two circuits and the danger of an overload on the neutral wire is avoided.

To connect the circuit breaker for the split-wired receptacle, place the two hot wires in the two breaker openings and tighten the screws. Run the grounding wire and the neutral wire to the neutral bar and fasten to separate terminals.

Figure 194 shows a split receptacle wired for two 120-volt circuits using two 12-2 w/g cables. When wired in this manner, you have two cables practically side by side from the SEP to the receptacle.

Figure 195 shows the same split receptacle wired with **one cable** (12-3 w/ground) and a **split 120/240-volt circuit.**

This is a less expensive way to wire the split receptacle. It saves both material and labor, since four wires will do the work of six, counting grounding wires. Proceed as follows to wire the split duplex receptacle using 12-3 w/g cable:

1. *Remove the break-off tab between the two brass terminals.*

   Assume you are installing a standard duplex receptacle with a break-off tab. Remove the tab between the **brass** terminals with pliers or screwdriver (Figure 191). Remember, the break-off tab between silver terminals is not removed, since one white neutral wire will serve both circuits.

2. *Connect the two hot wires (Figure 195).*

   Assume the receptacle box has a 12-3 w/g cable installed. One hot wire is red and one is black. The neutral is white and the grounding wire is bare.

   Connect the red wire to one brass terminal and the black wire to the other brass terminal. Where several duplex receptacles are to be wired, as over the kitchen counter, it is best to place the top receptacle on one circuit, for example, red, and the bottom receptacle on the other.

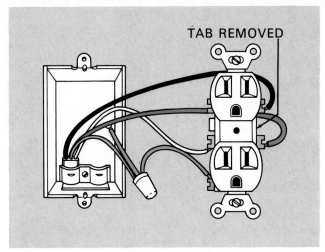

FIGURE 195. Split receptacle wired with 12-3 with ground cable.

3. *Connect the white neutral wire.*

   Connect the white neutral wire to one of the silver terminals. This neutral serves both hot wires. The tab should still be in place between the silver terminals.

4. *Connect the bare grounding wire.*

   Place the jumper wire ends from receptacle and box together with the end of the bare grounding wire from the cable. Fasten with a wire nut (Figure 196).

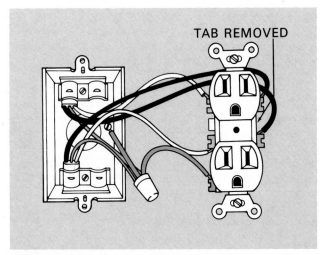

FIGURE 194. Split receptacle wired with two 12-2 with ground cables.

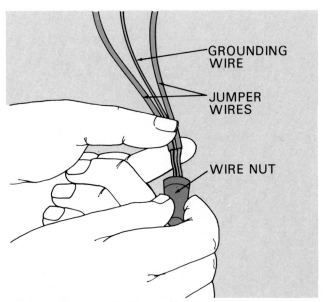

FIGURE 196. Connect grounding wire from cable to box grounding wire.

c. Connecting Individual Appliance Circuits

Appliances may be classified in three groups. They are:
- Portable.
- Stationary.
- Fixed.
- **PORTABLE APPLIANCES** such as coffee-makers, mixers and blenders may be plugged into the 20-ampere, 120-volt, kitchen, dining room and den circuits. They do not require special wiring and may be disconnected by pulling the plug from the receptacle.

In this section you are most concerned with stationary and fixed appliances which are wired on individual circuits.
- **STATIONARY APPLIANCES** are those that the homeowner usually moves with him from one house to another. Examples are self-contained ranges, clothes dryers and window air conditioners.
- **FIXED APPLIANCES** are those that are permanently installed such as built-in ovens and counter cooking tops, trash compactors, garbage disposals and water heaters.

Stationary and fixed appliances are generally wired on separate circuits. The Code requires that each appliance be protected by a **circuit breaker** at the SEP, or, if fuses are in a **pullout block,** no other disconnect is required. However, if the appliance is connected to a SEP containing only fuses, a separate disconnecting means is required. For example, the water heater is usually wired directly to connections at the tank. If you have a fused SEP without a pullout block for the water heater, a separate disconnect switch must be installed (Figure 197). If connected to a circuit breaker in the SEP, the special disconnect is not required.

Any water heater with capacity of 120 gallons or less must have fuse or circuit breaker protection at least 125% of the nameplate current. For example, a water

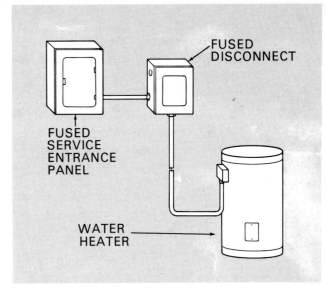

FIGURE 197. Disconnect required between a SEP with fuses and without a pullout block for the water heater.

heater with a nameplate rating of 30 amperes would require 40-ampere protection. (30 × 1.25 = 37.5 Install next highest common size, 40 ampere).

Table VIII shows typical wire sizes and circuit breaker or fuse sizes for home appliances and equipment. Check your local code and select the correct materials for appliances.

Connecting individual appliance circuits is discussed as follows:

(1) Connecting Self-Contained Range Receptacles.
(2) Connecting Separate Ovens and Countertop Cooking Units.
(3) Connecting Clothes Dryers.
(4) Connecting Water Heaters.
(5) Connecting Clothes Washers.
(6) Connecting Electric Heaters.

Load	Typical Wattage	Volts	Wire	Circuit Breaker Rating**	Comments
KITCHEN					
Range	12000	120/240	3 No. 6	50A-2P	Direct to Service Entrance
Range (double oven)	21000	120/240	3 No. 4	70A-2P	Direct to Service Entrance
Oven (built-in)	5000	120/240	3 No. 10	30A-2P	Direct to Service Entrance
Range Top	6800	120/240	3 No. 8	40A-2P	Direct to Service Entrance
Dishwasher	1200	120	2 No. 12	20A-1P	Recommend Separate Circuit*
Waste Disposer	300	120	2 No. 12	20A-1P	May be combined with Dishwasher
Broiler-Roaster	1500	120	2 No. 12	20A-1P	Kitchen Receptacle Circuits
Fryer	1300	120	2 No. 12	20A-1P	Kitchen Receptacle Circuits
Coffeemaker	1100	120	2 No. 12	20A-1P	Kitchen Receptacle Circuits
Toaster	1100	120	2 No. 12	20A-1P	Kitchen Receptacle Circuits
Grill—(Portable)	1000	120	2 No. 12	20A-1P	Kitchen Receptacle Circuits
Exhaust Fan	300	120	2 No. 12	20A-1P	Combine with Light Circuits
Kitchen Receptacles	1920	120	2 No. 12	20A-1P	Minimum of two circuits Consider 120/240 V.—3 wire Split Receptacle Circuit
Refrigerator	300	120	2 No. 12	20A-1P	Recommend separate circuit for
Freezer	350	120	2 No. 12	20A-1P	Refrigerator and Freezer
LAUNDRY					
Washer	1200	120	2 No. 12	20A-1P	Separate Laundry Circuit
Dryer	5000	120/240	3 No. 10	30A-2P	Direct to Service Entrance
Dryer (High Speed)	9000	120/240	3 No. 6	50A-2P	Direct to Service Entrance
Hand Iron	1000	120	2 No. 12	20A-1P	Use Laundry Circuit
Water Heater	5000	240	2 No. 10	30-40 2P	Consult Utility

*May be combined with waste disposer
**1P = 1-pole, 2P = 2-pole

(1) CONNECTING SELF-CONTAINED RANGE RECEPTACLES.
Assume you are wiring a flush mounted 50-ampere 125/250-volt receptacle (No. 10-50R, Table VII) in the kitchen. No. 6, 3-wire cable is installed in the receptacle box. For ranges of 8¼ KW or more rating, the minimum branch circuit rating must be 40 amperes to meet Code requirements.

The range operates at 120 volts on low heat and on 240 volts at higher settings. It requires a cable with three wires. The **white neutral wire** in a 3-wire cable may be used as a **ground wire** because of an exception in the Code. The exception is allowed for **ranges and clothes dryers only.** However, the exception allowed in NEC 250-60 does not permit the grounded circuit conductor (NEUTRAL) to be bonded to the frame of ranges and dryers in manufactured housing (mobile homes). Some jurisdictions will not allow this type installation in conventional construction. Also, the same section of the Code allows the use of service entrance cable with a bare neutral wire for grounding if the range or dryer is on a separate circuit and it originates at the SEP. Proceed as follows to connect a range receptacle:

1. *Remove the receptacle cover if required. Some receptacles do not require this step.*
2. *Note the terminal markings on the receptacle (Figure 198) and make the connections.*

Attach the hot wires (usually black and red) to the terminals marked "y" and "x." Connect the white wire to the terminal marked "W."

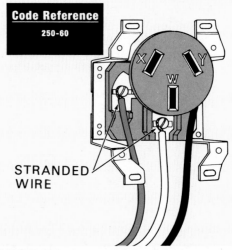

FIGURE 198. Connect the hot wires to terminals marked "x" and "y". Connect the white wire to the "W" terminal.

3. *Mount the receptacle in the box.*

NOTE: The grounding wire from the frame of the range connects to the grounded terminal of the receptacle when the connection (cord) from the range is plugged into the receptacle.

(2) CONNECTING SEPARATE OVENS AND COUNTERTOP COOKING UNITS. Except for different wire size, follow the same procedures to connect separate ovens and countertop cooking units as were used for connecting the self-contained range (Figure 198). Use the wire sizes shown in Table VIII or as recommended by the manufacturer.

NOTE: Receptacles are not required for the connection of separate ovens or countertop cooking units. They may be directly connected by bringing a properly sized cable from the SEP to the unit if a disconnect is provided in the SEP. The disconnect can be a circuit breaker or pullout fuse block.

(3) CONNECTING CLOTHES DRYERS. Procedures for connecting 240-volt clothes dryers are the same as for ranges. The wire size depends on the type of dryer. Most dryers are wired with No. 10 wire in a 30-amp circuit. If the dryer is a **high speed type,** one cycle may operate at up to 9000 watts. If you have this type, you must install a 50-amp receptacle and use No. 6 wire.

Proceed as for connecting ranges (Figure 198).

(4) CONNECTING WATER HEATERS. The water heater operates at 240 volts. It may require either No. 10-2 w/g for a 30-amp circuit or 12-2 w/g for a 20-ampere circuit, depending on local codes. Wiring is usually direct to the conductors in the water heater. However, some local codes may require a "junction"

(outlet) box located 30 or more centimeters (12 or more inches) above the heater. Procedures are given as follows:

Insert the cable. Connect as in Figure 200.

The cable contains one black wire, one white wire and a bare grounding wire. The white wire must be used as a hot wire to provide a 240-volt circuit. The Code requires the white wire to be identified as hot wherever it is accessible. This may be done by wrapping it with black tape at the splice in the box and at the breaker terminal in the SEP. Black paint may be used if preferred (Figure 200).

(b) If the **junction box is not required,** make the connections directly to the conductors in the water heater and connect the grounding wire to the metal case (Figure 201).

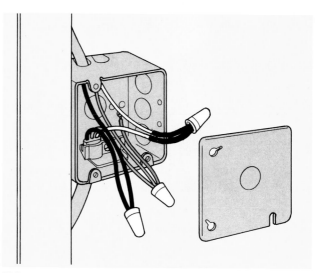

FIGURE 200. Install the cable and make required connections in the box.

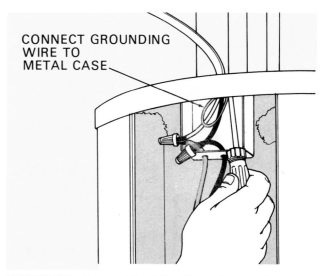

FIGURE 201. Connect the 240 volt conductors directly to heater conductors and the grounding wire to the metal case of the heater.

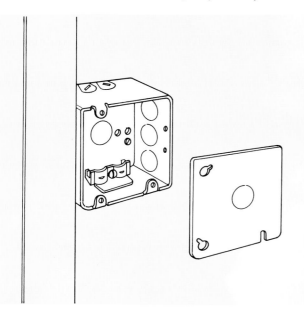

FIGURE 199. Junction box used only for making wiring connections. No device is installed.

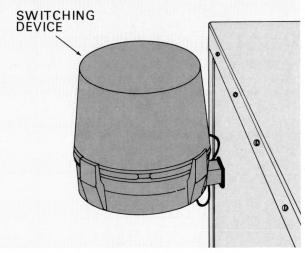

FIGURE 202. Switching device installed by power supplier to cycle the operation of water heaters and air conditioners during peak demand periods.

FIGURE 203. Rough-in wiring for central heating and air conditioning provided by the electrician.

In the interest of energy conservation, some electric power suppliers offer the homeowner lower rates for off-peak electrical consumption. A switching device installed by the power supplier controls the time-of-day operation of the heater. The device allows the heater to operate during the periods when demand for electricity on the lines is lower. Other types may cycle the heater operation, turning off electricity for a few minutes each hour. Power supplier personnel will connect this type of device (Figure 202).

(5) CONNECTING CLOTHES WASHERS. Most clothes washers can be plugged into a 120-volt receptacle and do not require special wiring. However, they should always be plugged into a receptacle on a separate circuit and be grounded.

(6) CONNECTING ELECTRIC HEATERS. Electric heat for residences includes sizes from electric furnaces large enough to heat an entire house, to small bathroom wall heaters. Where electric heat is installed, central whole house heating with electric furnace or heat pump is a common choice. However, a large number of individual room units of various kinds are installed. One example is the electric baseboard heater. They may be either 120-volt or 240-volt units. Baseboard heaters may have a built-in thermostat or may be controlled by a wall-type thermostat.

In most central heating installations, the electrician will provide the rough-in wiring (Figure 203) and the installer of the equipment will make the connections.

The electrician may be required to install electric baseboard heaters, wall heaters and ceiling bathroom heaters. Proceed as follows:

1. *Determine type and size branch circuit required.*

   Check the unit to determine whether a 120-volt or 240-volt circuit is required. You must also know the load demand of the heating unit to determine the correct wire size to install.

2. *Install cable if required.*

   Most heaters are installed on individual branch circuits. Run the cable to the outlet if not already installed. The branch circuit wires must have an ampacity of not less than 1.25 times the total ampere load of the heater (and fan, if any).

3. *Connect the heater.*

   (a) If you are installing a **ceiling heater,** as often used in bathrooms, mount a standard outlet box. Install the heater according to the instructions and connect the wiring. Ceiling heaters are usually prewired. Connect black to black and white to white wires or follow manufacturer's instructions. Install grounding wires as usual. (Instructions for installing switches are given in Section E. "Connecting Switches and Circuits.")

   (b) If you are installing a **baseboard heater** or a wall heater (Figure 204), an outlet box is not normally required. Install the heater on

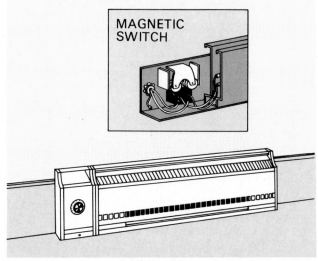

FIGURE 204. Wiring connections for baseboard heater.

the wall according to instructions included with the unit. Install a connector clamp through the opening in the heater box and make the connections to terminals on the unit or to wires attached to the unit (Figure 204).

d. Installing Junction Boxes

The junction box shown on page 89 illustrates one of its uses. A junction box may also be useful for adding outlets to an existing circuit. One or more cables may be connected in the box to new outlets if the circuit capacity is not exceeded (Figure 204A).

Proceed as follows:

1. *Turn off the power at the SEP and lock the cover.*

2. *Select the location for the junction box.*

    The location should be as near the new outlet as practicable and must be permanently accessible.

3. *Install the outlet box. Cut the cable on the selected circuit and install a square outlet box of the correct size for the number of wires to be connected.*

4. *Install one or more cables as needed for additional outlets.*

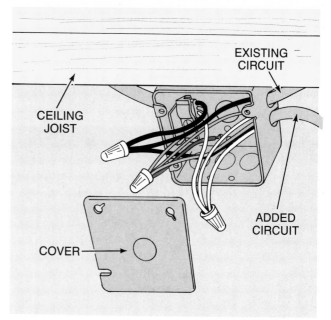

FIGURE 204A. A circuit may be added in a junction box if the circuit capacity is not exceeded.

5. *Complete wiring connections in the junction box. For each cable, connect black to black and white to white wires. Be sure to complete the grounding connections and install the cover.*

6. *Extend cable to the new outlets and make the required connections.*

D. Installing Lighting Fixtures

In this section, you will mount the lighting fixtures on walls and ceilings. Wiring connections will be made to the fixtures only. Switching connections are covered in detail in Section E.

Lighting fixtures are available in many sizes, styles and shapes. As installed in the home, they are either ceiling or wall fixtures. However, of more interest to you during installation is the method of mounting for the different types of fixtures. The following four mounting methods cover most installation procedures:

1. Mounting Direct-to-Box.
2. Mounting Strap-to-Box.
3. Mounting Fixture-to-Stud.
4. Mounting Recessed Fixtures.

1. MOUNTING DIRECT-TO-BOX

Most of the lighting fixtures you will install will have fixture wiring factory installed. The fixture wires are connected to the wires from the circuit cable (source) by wire nuts. However, the fixture mounted direct-to-box may have either lead wires or two terminal screws. This type of fixture with a wall switched outlet may be installed in a bedroom closet as in bedroom No. 1 in the standard house plan, provided the requirements of Code Section 410-8 are met. Additional information about fixtures in closets can be found on page 31 of this manual. This is a simple connection. Since it is at the end of a circuit, it requires only two terminals and a grounding connection. Fixtures with exposed metal now must have a grounding terminal or screw for the bare grounding wire connection.

fluorescent fixture must have at least six inches clearance horizontally between the fixture and the shelf.

1. *Connect circuit cable wires to fixture.*

   Attach the black wire to the brass screw, white wire to the silver screw and the bare ground to the box grounding screw (Figure 205).

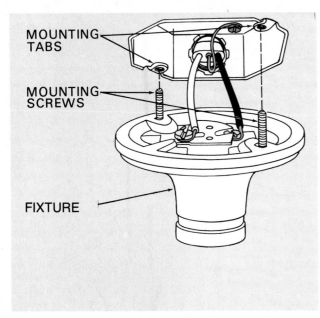

FIGURE 205. Fixture mounted directly to the box.

2. *Attach fixture to box.*

   Fold wires slightly and place fixture against the bottom of the box. Insert bolts and tighten (Figure 205).

   If the fixture is porcelain, do not overtighten as you may break the porcelain.

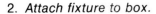

2. MOUNTING STRAP-TO-BOX

Assume you are mounting a wall fixture in a box in bathroom No. 1 of the standard house plan. Wires from the incoming cable (source) and from the outgoing cable (to following outlet) must be connected to the fixture wires. Proceed as follows:

1. *Connect circuit cable wires to fixture wires.*

   Connect three black wires with wire nut. Connect one each from **source cable** (top) from **fixture** (right) and from **outgoing cable** (bottom) (Figure 206). Connect three white wires with wire nut, one each from source, fixture and outgoing cable. Connect three bare grounding wires, one from each cable, one from the fixture, plus pigtail wire to grounding screw (4 wire ends total) (Figure 206). Fold wires into the box.

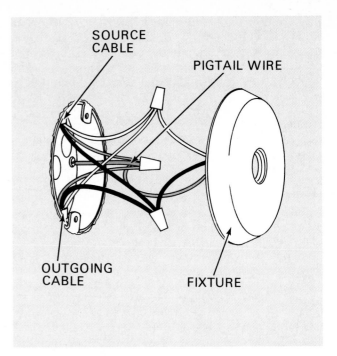

FIGURE 206. Connect wires from source cable, outgoing cable and fixture.

2. *Attach fixture to box.*

   Bolt strap to box and insert headless bolts. Attach fixture to headless bolts using ornamental nuts (Figure 207).

3. *If you have a ceiling fixture with bolts, attach to strap, as in Figure 207.*

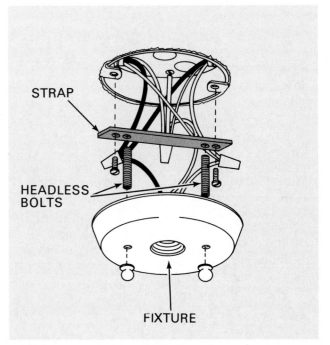

FIGURE 207. Attach strap to box and fixture to headless bolts.

92

3. MOUNTING FIXTURE-TO-STUD

Heavy suspended fixtures, such as large chandeliers, require more support than those described above. Many fixtures of this type require a **stem** in addition to the stud. The stud and stem are joined together by a fixture coupling sometimes called a hickey. (Figure 208). Proceed as follows:

1. *Screw coupling onto stud on outlet box.*

   Pull wires from fixture stem through coupling (Figure 208).

2. *Screw fixture stem into coupling and connect wires (Figure 208).*

3. *Push canopy against ceiling and tighten holding screw (Figure 208).*

A Code rule requires that lighting fixtures with exposed metal parts have a grounding terminal. You must connect the grounding wire to the terminal if you are using nonmetallic cable (Figure 208).

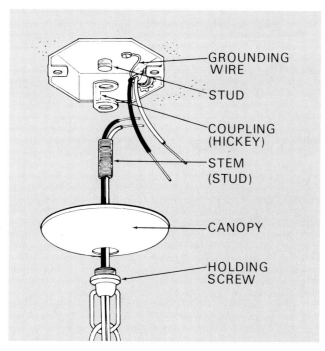

FIGURE 208. Screw coupling to stud and stem, connect wiring and attach canopy.

4. MOUNTING RECESSED FIXTURES

Recessed fixtures are self-contained. They should be mounted before plaster or dry wall is in place. Most are prewired and can be quickly mounted using adjustable brackets (Figure 209). Wiring connections are in the attached junction boxes. The Code requires that no thermal insulation may be installed within 8 cm (3 inches) of the recessed fixture enclosure, wiring compartment or ballast. Free circulation of air must be allowed around the fixture. Also, recessed portions of the enclosure must be spaced at least 1.3 cm (½ inch) from combustible material.

Proceed as follows:

1. *Mount the fixture according to procedures in Section II, "D. Installing Device Boxes and Outlet Boxes."*

2. *Connect wiring as required.*

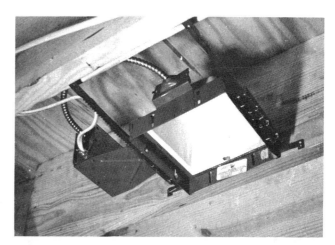

FIGURE 209. Recessed fixtures are prewired and are mounted with screws or nails. They must have 1.3 cm (½ inch) clearance between the fixture and the joist.

93

E. Connecting Switches and Circuits

The information contained in this section on connecting switches and circuits is very important in your training program. Upon successful completion of procedures included in the material, you will be able to connect switches to fit most situations you will find in common wiring jobs.

Connecting switches and circuits is discussed under the following headings:

1. Types and Sizes of Switches.
2. Connecting Single-Pole Switches.
3. Connecting Switched-Duplex Receptacles.
4. Connecting Three-Way Switches.
5. Connecting Four-Way Switches.
6. Connecting Combination Devices.
7. Connecting Double-Pole Switches.

> **NOTE: Before installing switches described in this section, be sure to read the installation instructions included in or on the carton.**

1. TYPES AND SIZES OF SWITCHES

In house wiring, the switches you install will likely be AC switches for control of lighting circuits. Most of those used in lighting circuits are toggle switches. Other kinds available are the rocker-type and push-button switches (Figure 210). Older type AC-DC switches are seldom used because the AC type is quieter and lower in cost.

Switches have their safe ampere and voltage ratings stamped on the metal yoke and, if horsepower rated, their maximum rating. A switch may be rated for both 120-volt and 240-volt operation. For example, 10 amperes at 120 volts or 5 amperes at 240 volts. In others the switch may be rated only for 120-volt use. Check your switches before mounting to be sure the ratings are suitable for the circuit.

Toggle Switch

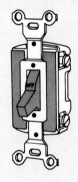

Rocker Switch
(DECORATOR STYLE)

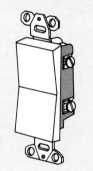

Push Button

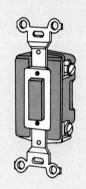

FIGURE 210. Types of switches used in the home.

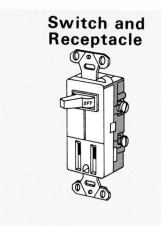

Switch and Receptacle

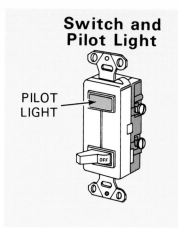

Switch and Pilot Light

PILOT LIGHT

Two Switches on one Strap

FIGURE 211. Examples of combination switches.

Switches are available in various combinations such as switch and receptacle, switch and pilot light or with two switches on one strap (Figure 211).

As with receptacles, switches may be either side-wired or back-wired or both. Other useful variations are the time-delay switch and the rotary-dimmer switch (Figure 212).

The **time-delay** switch turns lights on as an ordinary switch. However, when turned to "delay," which is the "off" position, the lamp remains "on" for a convenient period of time. In the garage in the standard house plan, for example, it will allow time to enter your car comfortably before the lamp turns "off."

A **rotary-dimmer** switch turns the lamp to any level of lighting desired, from "off" to full brightness. Dimmer switches are often used to control lighting levels in living and dining rooms.

Both time-delay and rotary-dimmer switches are mounted in standard switch boxes. Both connect in the circuit in the same manner as standard single-pole switches.

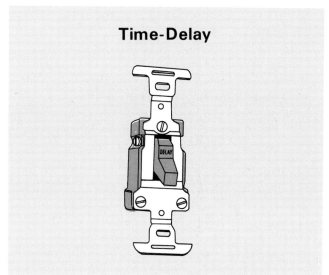

Time-Delay

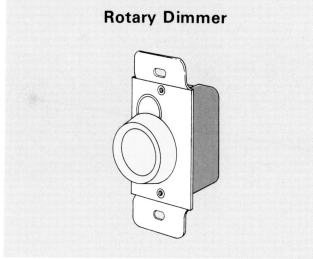

Rotary Dimmer

FIGURE 212. Time-delay and dimmer switches.

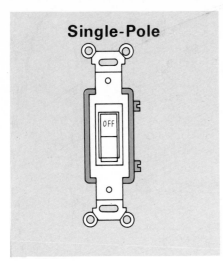 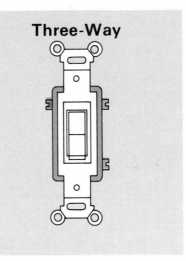 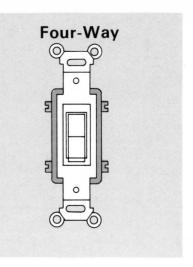

Single-Pole **Three-Way** **Four-Way**

FIGURE 213. **Single-pole, three-way and four-way switches.**

Switches also differ in their internal wiring. You are likely to work with three different types in house wiring. Terminals may be located on ends, front or sides. The three types are (Figure 213):

— Single-Pole Switches
— Three-Way Switches.
— Four-Way Switches.

2. CONNECTING SINGLE-POLE SWITCHES

A single-pole switch controls lighting from one position. But one single-pole switch may control several lights or receptacles on the same circuit. Wiring connections could be different for one or two or more lights on a circuit. You will also find different connections depending on whether the power source enters the circuit at the switch or through an outlet. The term **source** is used to indicate the power source from the SEP.

The Code requires that metal switch **boxes** be grounded. But the switches themselves do not usually have a grounding terminal unless used for a special purpose. So, unlike receptacles, the pigtail grounding wire is not normally required to be connected from box to switch.

A jumper wire is required to connect the grounding wire at the lighting outlet box, **only if the circuit continues to another outlet.** If the outlet is at the end of the circuit, the bare wire connects directly to the grounding screw in the outlet box. Make all required connections at each outlet or switch before moving to another.

Both of the above grounding situations are seen in the circuit in bedroom No. 1 in the standard house

plan (Figure 214). A single-pole switch controls the light. Current from the SEP enters at the switch box (source). The bare grounding wire is connected to the grounding screw in the switchbox by a jumper wire. The ends of the bare wire from the cables are connected to the jumper wire with a wire nut.

At the outlet box for the light, the bare grounding wire is connected directly to the grounding screw on the metal box. No jumper is required because the circuit does not continue beyond the lighting outlet.

Connecting single-pole switches is discussed as follows:

a. Connecting Single-Pole Switch with Switch Source.
b. Connecting Single-Pole Switch with Lighting Outlet Source.
c. Connecting Single-Pole Switch for Two Lamps Using Lighting Outlet Source.
d. Connecting Single-Pole Switch for Two Lamps with Switch Source and Unswitched Line.

a. Connecting Single-Pole Switch with Switch Source

Single-pole switches control lights from one position, as in a bedroom. The lever is marked "off" and "on" (Figure 213).

1. Connect switchbox wiring (bedroom No. 1, circuit No. 2) in standard house plan (Fig. 214).
   (1) Connect the black wire from the source cable to one terminal on the switch.
   (2) Connect the black wire from the outgoing cable to the other switch terminal.

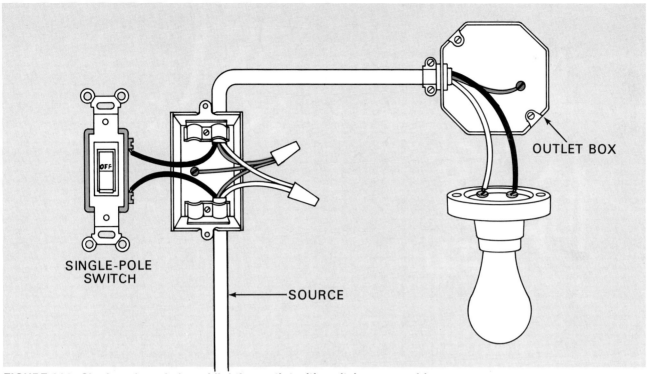

FIGURE 214. Single-pole switch and lighting outlet with switch-source wiring.

(3) *Connect the white neutral wire from each cable with a wire nut.*

Note that the white neutral wire does not connect to the switch.**Neutral wires do not connect to switches.**

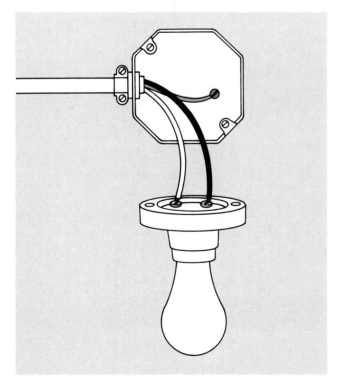

FIGURE 215. Outlet-box wiring for single-pole switch control.

(4) *Connect the bare grounding wire.*

Connect the jumper wire to the screw in the switch box. Place the ends of the bare wire from each cable together with the jumper wire and fasten with a wire nut.

2. *Connect outlet box wiring (Figure 215).*

(1) *Connect the grounding wire to the grounding screw in the outlet box.*

(2) *Connect cable wires to fixture terminals.*

Connect black wire to brass and white wire to silver terminal.

To simplify illustrations, bare grounding wire connections are not shown in some of the remaining circuits. You may refer to Figures 214 or 216A if necessary for bare grounding wire connections. If the circuit continues beyond the outlet box at the fixture, connect the grounding wires from cable ends to the box by means of a jumper wire. Use the procedure shown at the switch box in Figure 216A.

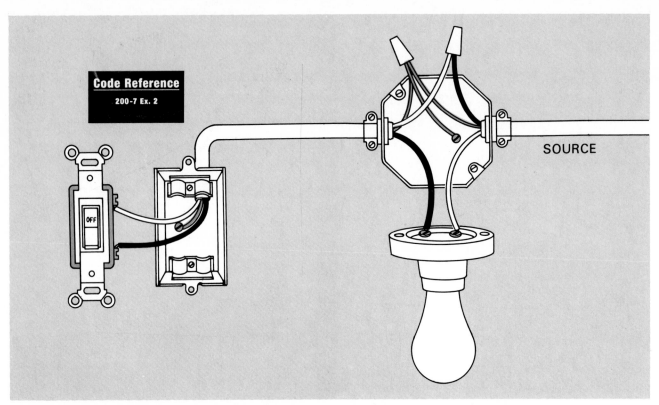

FIGURE 216A. Where the white wire is installed on the supply side of the switch loop, it is not necessary to identify as a hot wire.

b. Connecting Single-Pole Switch with Lighting Outlet Source

You may want to connect either (1) one single-pole switch to one light outlet or (2) two single-pole switches to two different light outlets.

(1) ONE SINGLE-POLE SWITCH TO ONE LIGHT OUTLET (Figure 216A). An example of the single-pole switch with lighting outlet source is in bedroom No. 3, circuit No. 3 of the standard house plan.

1. *Connect wiring at the light outlet.*

   (1) *Connect white wire from the switch to black wire from source cable.*
   Remember that an **exception** was noted in the first discussion of Code requirements regarding the **white wire** in a cable (Section B, 1, "Grounding the System and Circuits"). The exception is as shown in a **switch loop**, sometimes called a switch leg. A switch loop is required when the power source enters the lighting outlet box first (see Code Section 200-7, Ex. 2).
   The Code permits connecting the white wire in a cable to a switch, **provided it is used for the supply to the switch but not as a return conductor from the switch to the switched outlet.**

In this instance and all others **where permitted, the white wire is used as if it were a black wire.** The reason it is permitted is that two-wire cable is made only with one black and one white wire, not two black wires. This white wire is considered an unidentified or "hot" wire.

To connect a single-pole switch with lighting outlet source, proceed as follows (Figure 216A):

(2) *Connect black wire from switch to brass terminal on the fixture.*

(3) *Connect white wire from source to silver terminal on fixture.*

(4) *Connect bare grounding wires.*
   Connect pigtail grounding wire to ground screw in switch box. Connect wire ends from pigtail wire, source cable and switch-box cable with wire nut.

2. *Connect wiring at the switch.*
   (1) *Connect the black wire to one switch terminal.*
   (2) *Connect the white wire to the other switch terminal.*
   (3) *Connect bare grounding wire to box grounding screw.*

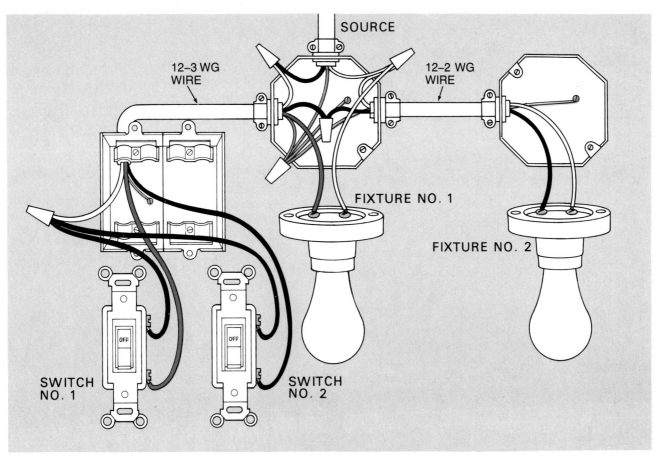

FIGURE 216B. Two single-pole switches control two different light outlets.

(2) TWO SINGLE-POLE SWITCHES TO TWO DIFFERENT LIGHT OUTLETS (Figure 216B). The grounding wires are not shown in this illustration so the other wiring can be shown more clearly. Connect grounding wires as in Figure 216A.

1. *Connect wiring at fixture No. 1.*

   This circuit requires 3-wire cable from the source to the switch and 2-wire to fixture No. 2.

   (1) *Connect black wire from source to white wire from switch box.*

   (2) *Connect white wire from source to a white jumper wire and to white wire going to fixture No. 2.*

   (3) *Connect black wire going to the switch to black wire from fixture No. 2.*

   (4) *Connect red wire to brass-colored light terminal and white jumper wire to silver light terminal.*

2. *Connect wiring at Fixture No. 2.*

   (1) *Connect black wire to brass terminal and white wire to silver terminal.*

3. *Connect wiring at switches.*

   (1) *Connect black wire from box at fixture No. 1 to a terminal on switch No. 2.*

   (2) *Connect white wire from box at fixture No. 1 to two black jumper wires.*

   (3) *Connect red wire from fixture No. 1 to a terminal on switch No. 1.*

   (4) *Connect one black jumper wire to Switch No. 1 and the other black jumper wire to switch No. 2.*

c. Connecting Single-Pole Switch for Two Lamps Using Lighting Outlet Source

Assume you are wiring the wall lighting outlets in bathroom No. 1 of the standard house plan on one switch. Grounding wire connections are omitted for greater clarity of circuit wire connections.

The circuit begins at the lighting fixture on the left. Connections are similar to those for Figure 216 except that the outgoing cable to the second outlet must be included.

Proceed as follows (Figure 217):

1. *Connect wiring at the switch.*

    (1) *Connect the black wire to one switch terminal.*

    (2) *Connect the white wire to the other switch terminal.*

    (3) *Connect bare grounding wires as in previous section (not shown).*

2. *Connect wiring at first outlet box.*

    (1) *Connect white wire from the switch to black wire from source cable.*

    (2) *Connect black wire from switch to black wires from fixture and cable going to next lighting outlet.*

    (3) *Connect white wire from source cable to white wires from fixture and cable going to next lighting outlet.*

    (4) *Connect grounding wire as in previous sections (not shown).*

3. *Connect wiring at second outlet box.*

    (1) *Connect black wire to brass terminal.*

    (2) *Connect white wire to silver terminal.*

    (3) *Connect grounding wire as in previous section (not shown).*

Circuit No. 1 ends at this outlet.

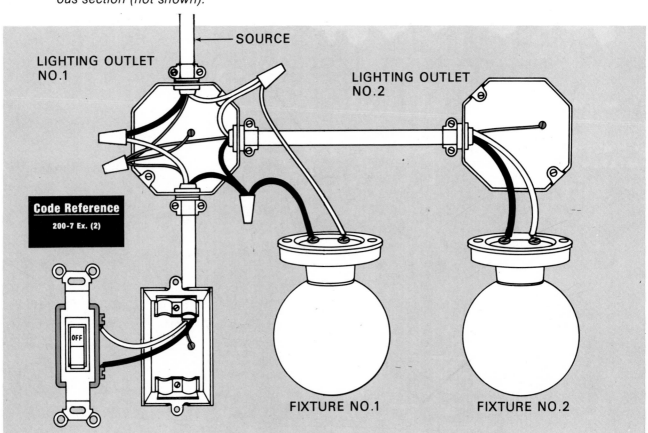

FIGURE 217. Single-pole switch controls two lighting outlets with lighting outlet source.

d. Connecting Single-Pole Switch for Two Lamps with Switch Source and Unswitched Line

In the standard house plan, the circuit often continues from outlet to outlet. **Three-wire** cable is required if an *unswitched* cable is to be continued beyond the light fixtures. The 3-wire cable contains a white, a black, and a red wire (Figure 218). The black and red wires are both hot. Figure 219 shows an example of a switch source circuit, two wall lamps with two lights and an unswitched cable continuing beyond the lights.

Proceed as follows using 14–3 w/g or 12–3 w/g cable (Figure 219).

1. *Connect wiring at the switch.*
   (1) *Connect white wire from source to white wire of outgoing cable.*
   (2) *Connect black wire from source to black wire of outgoing cable and to 15 cm (6-inch) black jumper. Connect black jumper to either switch terminal.*
   (3) *Connect red wire of outgoing cable to other switch terminal.*
   (4) *Connect bare grounding wires as in previous section (not shown).*
2. *Connect wiring at first outlet box.*
   (1) *Connect white wire from switch to white wire in outgoing cable and white wire from fixture.*
   (2) *Connect red wire to red wire in outgoing cable and to black wire from fixture.*
   (3) *Connect black wire to black wire.*

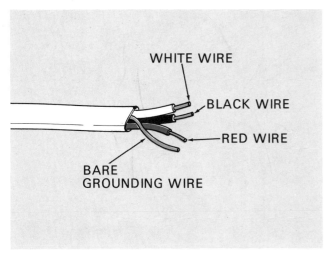

FIGURE 218. Three-wire with ground cable.

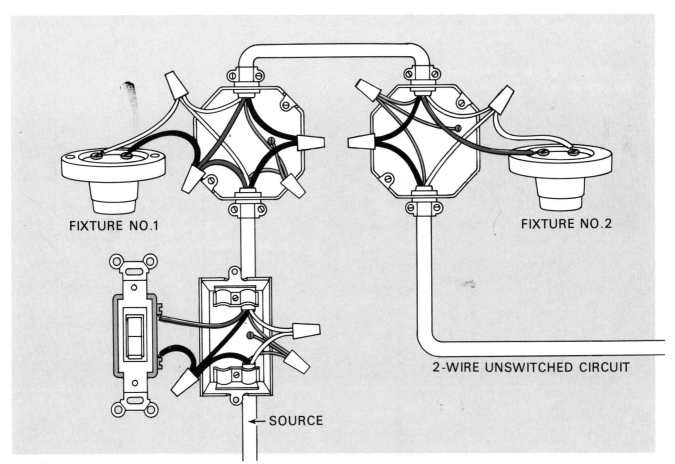

FIGURE 219. Single-pole switch with two lamps and unswitched line.

(4) *Connect grounding wires as in previous sections (not shown).*

3. *Connect wiring at second outlet box.*

(1) *Connect white wire to white wire from both cables and to silver terminal on fixture.*

(2) *Connect black wire to black wire from each cable.*

(3) *Connect red wire from first outlet to bronze terminal on fixture.*

(4) *Connect grounding wires as in previous section (not shown).*

As shown (Figure 219), a two-wire un-switched circuit extends beyond the second lighting outlet to the convenience outlet. If desired, one or more additional light outlets could be added, using 3-wire cable from outlet to outlet, until the point where the unswitched line is needed.

3. CONNECTING SWITCHED DUPLEX RECEPTACLES

Connecting switched-duplex receptacles is discussed as follows:

a. Connecting Switched-Duplex Receptacles.

b. Connecting Switched-Split-Duplex Receptacles.

a. Connecting Switched-Duplex Receptacles

Switched-duplex receptacles are often installed in living areas of the home. Table lamps supplying part, or all of the lighting, are more convenient to use if connected to a switched receptacle that you can turn on when entering a room. The switch may control both outlets on the receptacle or only one.

To connect the switch to both halves of a duplex receptacle outlet, proceed as follows:

Assume you are wiring a bedroom circuit in the standard house plan. The source is at a wall receptacle. The switch is near the bedroom door. It controls certain bedroom receptacles (Figure 220).

1. *Connect wiring at the receptacle.*

Connect black wire from switch to either brass terminal. Connect white wire from switch to black wire from source cable. Connect white wire from source cable to silver terminal on receptacle.

2. *Connect cable going to next receptacle.*

Connect black wire to brass terminal and white wire to silver terminal.

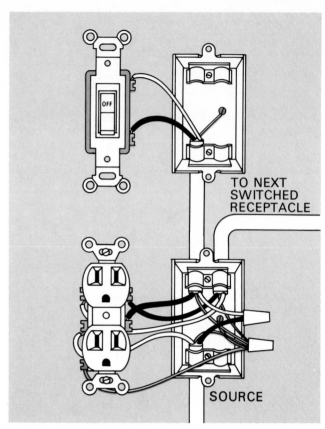

FIGURE 220. Wiring for a switched-duplex receptacle.

TO NEXT SWITCHED RECEPTACLE

SOURCE

3. *Connect grounding wires (not shown).*

Connect grounding wires as in previous sections.

4. *Connect wiring at the switch.*

Connect the black wire to one terminal and white wire to the other terminal.

b. Connecting Switched-Split-Duplex Receptacles

It is often desirable to have one half of the receptacle controlled by a switch and have the other half remain constantly hot. In the standard house plan, assume you are wiring the living room receptacle on the wall next to the bedroom with switch in the hall.

Proceed as follows:

1. *Remove connecting tab between two brass terminals of the receptacle.*

Use screwdriver or pliers to remove tab. The outlets are no longer electrically connected on the hot side when tab is removed.

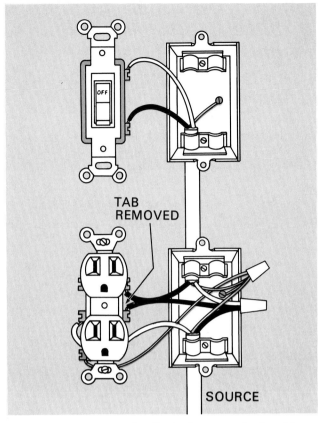

FIGURE 221. Switched-split-duplex receptacle wiring.

2. *Connect wiring at receptacle (Figure 221).*

   Procedures are the same as for a single circuit except for one black jumper wire to a brass terminal. Connect black source wire to a black jumper and to white wire from switch. Connect black jumper to lower brass terminal. The jumper wire makes the lower outlet constantly hot (not controlled by switch). Connect black wire from switch to top brass terminal. This outlet is controlled by the switch. Connect source white wire to silver terminal.

3. *Connect wiring at the switch.*

   Connect black wire to one terminal and white wire to other terminal (Figure 221). Connect grounding wire to grounding screw in box (not shown).

4. *Connect grounding wires as in previous sections.*

---

4. CONNECTING THREE-WAY SWITCHES

Three-way switches make it possible to switch lighting fixtures on or off from either of **two** locations. Why not three locations if they are three-way switches? The three-way description refers to the **number of terminals** on the switch, not the number of locations. Three-way switches have three terminals instead of two, as on single-pole switches. With the use of three-way switches, it is convenient to turn lights on or off without retracing steps. You may turn the light on at the switch nearest the door where you enter the room and turn it off at the other switch as you leave (Figure 222). Three-way switches are also useful in long halls, stairways and between house and garage.

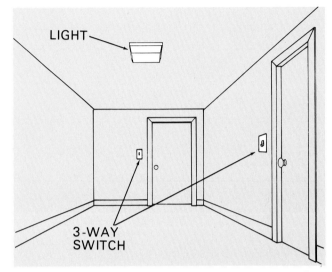

FIGURE 222. Three-way switches turn light on or off at two locations.

Three-way switches are used only in pairs. The toggle (handle) on a three-way switch is not marked for "on-off" position (Figure 223). Either up or down position may switch the light on or off, depending on the position of the toggle on the other switch.

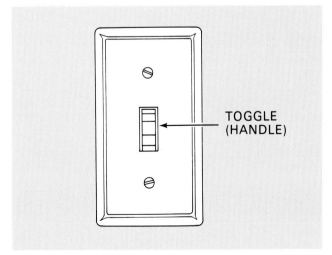

FIGURE 223. The toggle on a three-way switch is not marked to indicate "on-off."

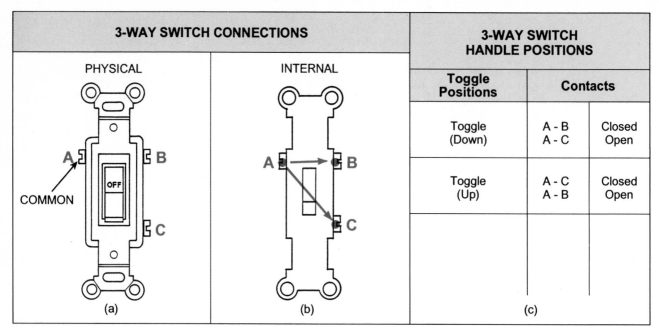

3-WAY SWITCH CONNECTIONS		3-WAY SWITCH HANDLE POSITIONS		
PHYSICAL	INTERNAL	Toggle Positions	Contacts	
(a)	(b)	Toggle (Down)	A - B A - C	Closed Open
		Toggle (Up)	A - C A - B	Closed Open
		(c)		

FIGURE 223A. A three-way switch as it appears (left) and as a schematic (right).

NOTE: It should be noted that there always exists a set of made contacts within the three-way switch, independent of whether the load is energizer or not. There is always a connection internally in the three-way switch from the common (hinge) to one of the available terminal outputs (Figure 223A).

The **power source** of the three-way switch circuits is **two-wire** cable (14–2 w/g or 12–2 w/g). For wiring *between* the three-way switches, you must use a minimum **three-wire cable** with ground.

On a three-way switch, one of the three terminals is a **common** or pivot terminal. You must identify the common terminal to wire the switch correctly. Usually it is not hard to locate. The common terminal may be a different color than the other two (Figure 224), usually darker. But on some switches the word "common" is printed next to the common terminal. It is often alone or on one side, but this varies according to the manufacturer. The carton or packing slip will indicate its location on the switch if not otherwise identified.

There are several possible combinations of lights and switches in 3-way switch wiring. As with single-pole switches, connections are different depending on where the power source enters the circuit. Also, the position of the light fixture in relation to the switches affects the wiring procedures.

The **power source** of the 3-way switch circuits is **two-wire** cable (14–2 w/g or 12–2 w/g). For wiring *between* the switches, you must use **three-wire cable**.

Connecting three-way switches is discussed as follows:

  a. Rules for Wiring Every Three-Way Switch.
  b. Wiring Three-Way Switches with Lighting Outlet Ahead of Switches Using Light Outlet Source.
  c. Wiring Three-Way Switches with Lighting Outlet Between Switches Using Switch Source.
  d. Wiring Three-Way Switches with Light at End of Run Using Switch Source.

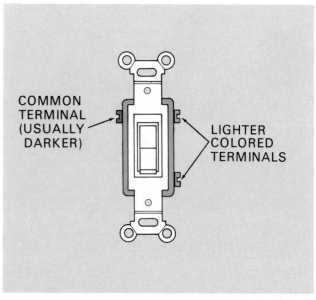

FIGURE 224. The common terminal screw on a three-way switch is usually, but not always, darker than the other two terminals.

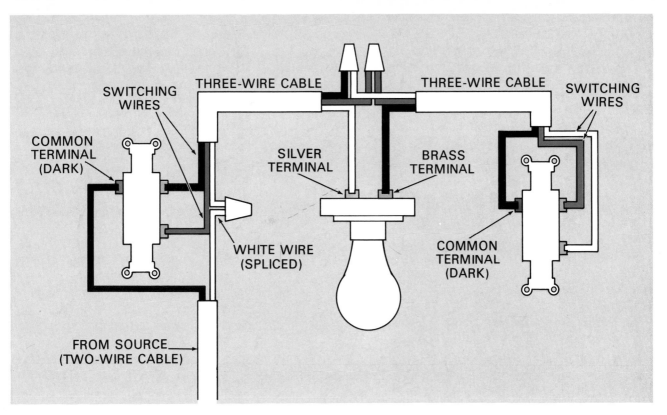

FIGURE 225. Basic wiring layout for wiring a three-way switch.

a. Rules for Wiring Every Three-Way Switch

Wiring procedures for three-way switch circuits can be confusing, especially in the beginning. However, there are rules for wiring these circuits that will simplify the wiring procedures if carefully observed. When wiring a three-way switch circuit using 2 and 3 wire cable, the rules are as follows (Figure 225).

> NOTE: Figure 225 is intended only to demonstrate the application of the basic rules shown below. Detailed procedures for actual wiring situations are shown in figures 226, 227 and 228.

1. *Connect white wire from source to silver terminal on the light fixture.*

   It will be spliced on some circuits but not connected at any other terminal. Note in figure 225, the white wire is shown as a spliced wire.

2. *Connect black wire from source to common terminal of either switch.*

3. *Connect black wire from brass terminal on light fixture to common terminal of the other switch.*

   These three steps complete the connection to the common terminals on each switch and to both terminals on the light fixture.

4. *To complete the circuit, connect the two remaining terminals of one switch to the two remaining terminals of the other switch.*

   The remaining wires (shown as a red wire and a black wire spliced to white) are known as **switching** or **traveler** wires. These wires always connect the remaining two lighter colored terminals of one switch to the remaining lighter colored terminals of the other switch.

b. Wiring Three-Way Switches with Lighting Outlet Ahead of Switches Using Light Outlet Source

Wiring from the source to the first switch box on the circuit is 14−2 w/g or 12−2 w/g cable (Figure 226). Three-wire w/g cable is used between the switches. Insulation on wires in three-wire cables is usually **black, white** and **red**. Proceed as follows:

1. *Connect wiring in lighting outlet box (Figure 226).*

   (1) *Connect white wire from the source cable to silver terminal on lighting fixture.*

   (2) *Connect black wire from common terminal of switch No. 1 to brass terminal on light fixture.*

   (3) *Connect black wire from source to white wire from switch No. 1.*

   This white wire is spliced at switch box No. 1 to the white wire in the three-wire cable and continues to switch No. 2 common terminal. As in the case of previous switched circuits, the white wire serves as a hot wire for the supply to switch. When used on the supply side of the switch circuit, the Code does not require it to be identified (painted black).

   (4) *Connect grounding wires as in previous sections.*

2. *Connect wiring to switch No. 1 and box (Figure 226).*

   (1) *Splice white wire from lighting outlet cable to white wire from cable to second switch (used as hot wire to switch No. 2).*

   (2) *Connect black wire from lighting outlet cable to common terminal on switch No. 1.*

   (3) *Connect black wire from three-wire cable to a light-colored terminal on switch No. 1 (switching wire).*

   (4) *Connect red wire from three-wire cable to other light-colored terminal (switching wire).*

   (5) *Connect grounding wires and ground the switch box.*

3. *Connect wiring to switch No. 2 and box (Figure 226).*

   (1) *Connect white wire to common terminal of switch No. 2.*

   (2) *Connect black wire to a light-colored terminal of switch No. 2.*

   (3) *Connect red wire to a light-colored terminal of switch No. 2.*

   (4) *Connect grounding wire.*

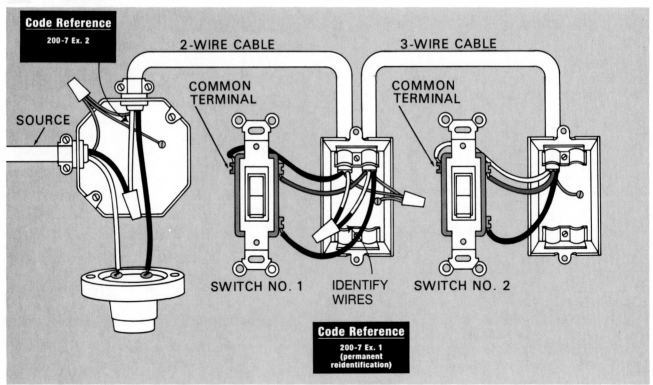

FIGURE 226. Wiring for three-way switch with lighting outlet source.

c. Wiring Three-Way Switches with Lighting Outlet Between Switches Using Switch Source

Assume you are wiring switches for a **hall light** (Figure 227):

1. *Connect wiring at switch No. 1.*
   (1) *Connect black wire from source cable to common terminal of switch.*
   (2) *Splice white wire from source cable to white wire from outgoing cable to lighting outlet box.*
   (3) *Connect black wire from outgoing cable to light-colored terminal on switch No. 1.*
   (4) *Connect red wire from outgoing cable to other light-colored terminal on switch No. 1.*
   (5) *Connect switch grounding wires (not shown).*

2. *Connect wiring to lighting outlet box (Figure 227).*
   (1) *Connect white wire from switch No. 1 cable to silver terminal on light fixture.*
   (2) *Splice black wire from switch No. 1 cable to white wire from switch No. 2 cable.*
   (3) *Splice red wires from each cable together.*
   (4) *Connect black wire from switch No. 2 cable to brass terminal on light fixture.*
   (5) *Connect grounding wires and bond box where required.*

3. *Connect wiring to switch No. 2 (Figure 227).*
   (1) *Connect black wire to common terminal.*
   (2) *Connect red wire to light-colored terminal.*
   (3) *Connect white wire to other light-colored terminal.*
   (4) *Connect switch grounding wire (not shown).*

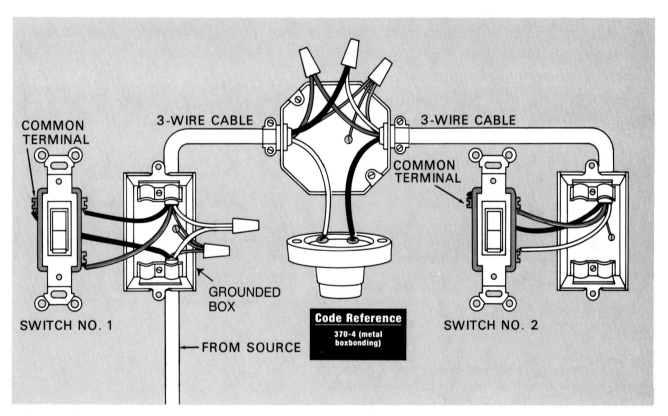

FIGURE 227. Wiring for three-way switches with lighting outlet between switches.

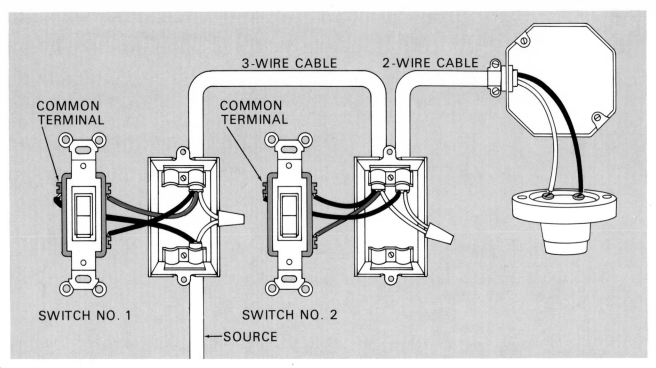

FIGURE 228. Wiring for three-way switches with light at end of run using switch source.

d. Wiring Three-Way Switches with Light at End of Run Using Switch Source

Assume you are wiring a lighting fixture with switches ahead of the light at the end of a run (Figure 228).

1. *Connect wiring in switch No. 1.*
   (1) *Connect black source wire to common terminal of switch No. 1 and white source wire to white wire in three-wire cable going to switch No. 2. Connect red and black wires (switching wires) from three-wire cable to light-colored terminals.*
   (2) *Connect grounding wires (not shown).*
2. *Connect wiring in switch No. 2 (Figure 228).*
   (1) *Connect black wire from outgoing cable to switch No. 2 cornmon terminal.*
   This wire also connects to brass terminal of fixture.
   (2) *Splice white wire to white wire.*
   (3) *Connect black wire and red wire from switch No. 1 to other two switch No. 2 terminals.*
   (4) *Connect grounding wires (not shown).*
3. *Connect wiring to lighting fixture (Figure 228).*
   (1) *Connect black wire to brass terminal and white wire to silver terminal.*
   (2) *Connect grounding wire to box (not shown).*

5. CONNECTING FOUR-WAY SWITCHES

Four-way switches are **used only in combination with two three-way switches** to control lighting from

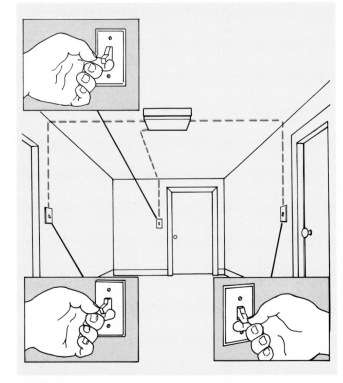

FIGURE 229. A circuit containing one four-way switch and two three-way switches allows control from three locations.

108

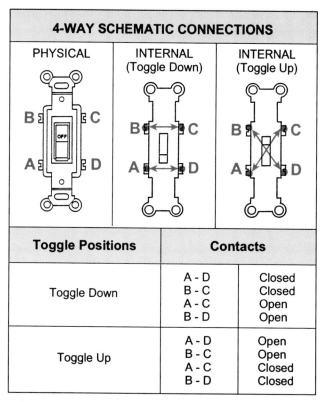

4-WAY SCHEMATIC CONNECTIONS

	PHYSICAL	INTERNAL (Toggle Down)	INTERNAL (Toggle Up)

Toggle Positions	Contacts	
Toggle Down	A - D	Closed
	B - C	Closed
	A - C	Open
	B - D	Open
Toggle Up	A - D	Open
	B - C	Open
	A - C	Closed
	B - D	Closed

FIGURE 229A. Physical and internal connections showing how a four-way switch works.

three or more locations. A circuit containing two three-way switches and one four-way switch gives control from three locations (Figure 229). By adding four-way switches to the circuit, you can have as many control locations as desired for the same lighting outlet(s). Four-way switches are so called because they have **four terminals**. Installation of four-way switches is easier than the three-way. You will use three-wire cable with ground between all switches and two-wire cable between the light and the first switch.

NOTE: It should be noted that there always exists a set a double-made contacts within the four-way switch, independent of whether the load is energizer or not. There is always a double-set connection internally in the four-way switch to the available terminal outputs (Figure 229A).

Assume you are installing the switches in the kitchen of the standard house plan. The source enters at the lighting outlet. Install 14–3 w/g cable between switches and 14–2 w/g cable between lighting outlet and first switch.

Steps for connecting four-way and three-way switches with source at the lighting outlet are as follows (Figure 230):

1. *At the fixture, connect the black wire from switch No. 1 to the brass terminal on the fixture. Connect the white wire from the source to the silver terminal on the fixture. Splice black*

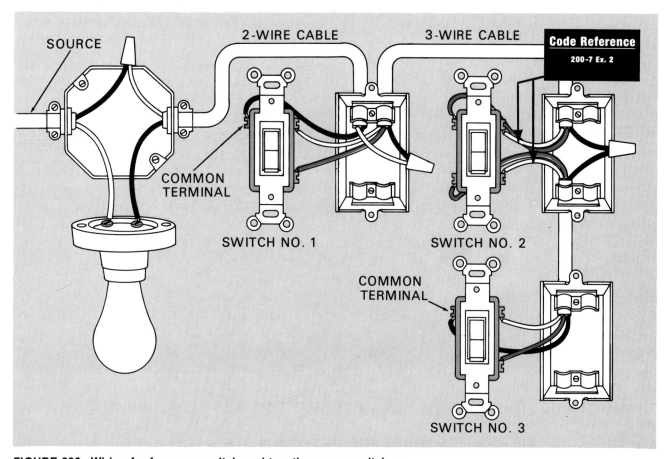

FIGURE 230. Wiring for four-way switch and two three-way switches.

109

wire from source to white wire going to the box at switch No. 1.

2. *At the box for switch No. 1, splice black wire from switch No. 2 to the white wire going toward the fixture. Connect white and red wires from switch No. 2 to the light-colored terminals. Connect black wire from light fixture to common terminal on switch No. 1.*

3. *At the box for switch No. 2, the four-way switch, splice together the black wires from incoming and outgoing cable. The black wire does not connect to switch No. 2. Connect the switching wires, white wires from both cables, to terminals on one side. Connect red wire from switch No. 3 to terminal opposite white wire from switch No. 3. Connect second red wire to remaining terminal.*

4. *At switch No. 3, connect the black wire to the common terminal, white wire to a light-colored terminal and red wire to the other light-colored terminal.*

5. *Connect grounding wires at each box (not shown).*

6. CONNECTING COMBINATION DEVICES

Connecting combination devices is discussed as follows:
a. Connecting Combination Switch and Receptacle.
b. Connecting Switch with Pilot (Locator) Light.

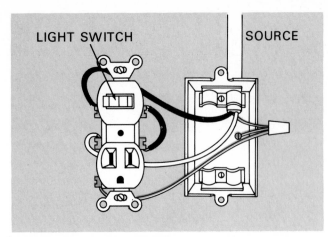

FIGURE 232. Switch controls only the receptacle outlet.

a. Connecting Combination Switch and Receptacle

Many options are available for wiring a combination switch and receptacle. Directions for wiring various kinds of circuits are usually given on the carton or on a packing slip included with combination devices. An example is seen in Figure 231.

In Figure 231 the switch is connected to a light only. It does not control the outlet. The receptacle is unswitched, always hot. In Figure 232 the switch controls only the receptacle. In Figure 233A, the switch and outlet are on split-wired 120/240-volt circuits. The switch controls a light and the receptacle is available for other use.

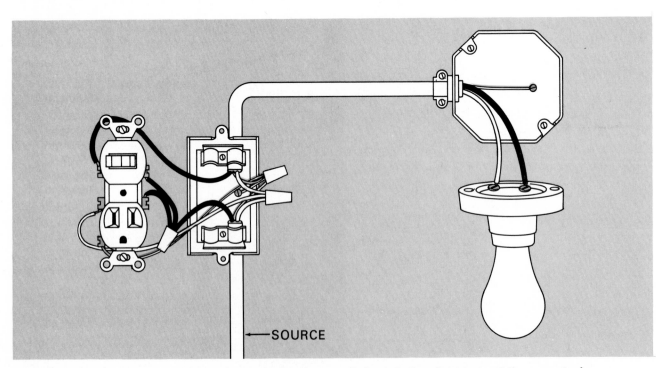

FIGURE 231. Combination switch and receptacle where switch controls a light but not the receptacle.

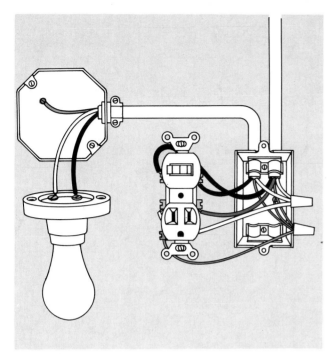

FIGURE 233A. The switch and outlet are on split-wired circuits.

glows when the switch is **off.** This locates the switch for you in the dark. When wired another way, the pilot light glows when the switch is **on.** This last wiring method is useful when a light or equipment is out of sight of the switch position. An example is a basement light controlled by a switch beyond the basement door. Connections are different for different manufacturers. Follow directions included with the pilot light.

7. CONNECTING DOUBLE-POLE SWITCHES

A 240-volt circuit has two hot wires. The switch for a 240-volt circuit must be a double-pole type so both hot wires will be broken when the switch is turned off. Wire the switch and receptacle connections as shown in Figure 233B. Wrap white wires with black tape at each terminal to show they are hot. This is classified as a multiwire circuit by the Code. Simultaneous disconnect of both legs of the circuit is required in the SEP circuit breaker for multiwire circuits.

b. Connecting Switch with Pilot (Locator) Light

A switch with pilot light may be connected either of two ways. When connected one way, the pilot light

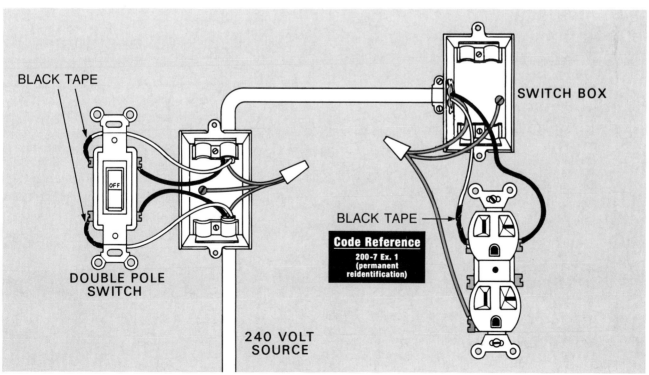

FIGURE 233B. Double-pole switch required for 240-volt circuit.

111

F. Installing Doorbells, Chimes and Smoke Detectors

Installing doorbells, chimes and smoke detectors is discussed under the following headings:

1. Installing Doorbells and Chimes.
2. Installing Smoke Detectors.

1. INSTALLING DOORBELLS AND CHIMES

In new homes, doorbells have largely been replaced by chimes. Most homes have pushbuttons at the front and rear doors and sometimes at a third door. The pushbuttons may each have a different note that indicates from which door the signal comes.

Doorbells and chimes operate at low voltage, usually 16 to 24 volts AC. A transformer must be installed to reduce the voltage as required. For doorbells and chimes, the transformer reduces the voltage from 120 volts to the 16 to 24 volts required. Number 18–2 or 20–2 insulated wire may be used for connections between the transformer, the chimes and pushbuttons. Assume you are installing a front and rear door chime system.

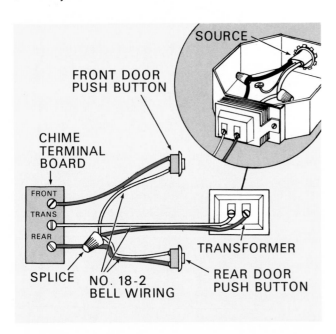

FRONT DOOR PUSH BUTTON

SOURCE

CHIME TERMINAL BOARD

FRONT

TRANS

REAR

SPLICE   NO. 18-2 BELL WIRING

TRANSFORMER

REAR DOOR PUSH BUTTON

FIGURE 234. Circuit wiring for a front and rear door (two-note) chime system.

Proceed as follows (Figure 234):

1. *Install the transformer.*

   Mount the transformer on an outlet box. Most will have a bracket that can be attached to a knockout in the box. Mount according to manufacturer's instructions. Make line voltage connections inside the box.

2. *Connect 120-volt circuit to transformer.*

   Connect white wire to silver terminal and black wire to brass terminal or white wire to white wire, and black wire to black wire. Connect the bare wire to a grounding screw or clip on the outlet box.

3. *Connect low voltage wiring pushbuttons and chimes (Figure 234).*

   (1) *Using 18–2 cable, connect red and white wires at transformer.*

   (2) *Connect white wire from transformer to center terminal on chime terminal board (staple wires for each run).*

   (3) *Connect red wire from front door button to front door terminal at the chime.*

   (4) *Connect red wire from rear door button to rear door terminal at the chime.*

   (5) *Splice white wires from each doorbell to red wire from transformer.*

2. INSTALLING SMOKE DETECTORS

Smoke detector units are now available that are sensitive to very small amounts of smoke and/or gases produced by combustion. An alarm sounds to warn of danger. These units are relatively low in cost. Many building codes require that they be installed in all new homes. Two or more units may be required in a 2-story house.

Installing smoke detectors is discussed as follows:

a. Installing Battery-Operated Smoke Detectors.
b. Installing 120-Volt Smoke Detectors.

a. Installing Battery-Operated Smoke Detectors.

Battery-operated types are self-contained and do not require any wiring. They may be mounted on the wall or ceiling. Follow instructions included with the product to select the best location. Fasten the unit to wall or ceiling as follows:

1. *Place template on mounting surface (Figure 235).*

   A template is usually supplied with unit. Tape the template on the ceiling or wall and drill holes for the plastic screw anchors if required. Tap the anchors into the holes with a hammer, then remove the template.

2. *Attach unit to wall or ceiling (Figure 235).*

   Insert screws through the base of the alarm unit and tighten them into the plastic anchors to complete the job.

   An alarm device warns when the battery needs to be replaced. In Figure 236, a small flag pops out. Other types have audio signals. A battery usually lasts about one year before requiring replacement.

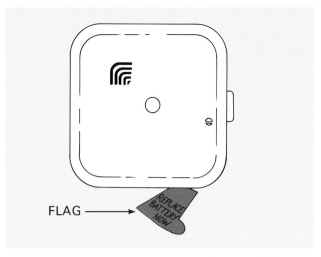

FLAG

FIGURE 236. Small flag pops out of unit to indicate a weak battery. Other types have audio signals.

b. Installing 120-Volt Smoke Detectors

Follow instructions of the manufacturer to locate the unit correctly. Remember, the circuit must have power at all times and cannot be switched off.

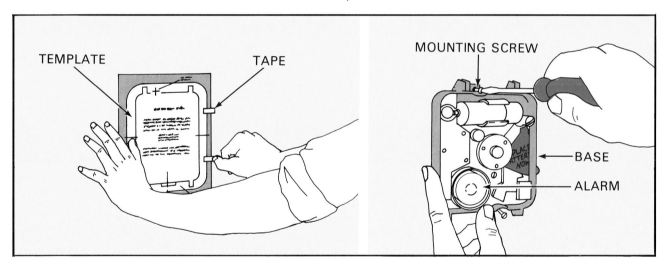

TEMPLATE   TAPE

MOUNTING SCREW

BASE

ALARM

FIGURE 235. Smoke detector attached to wall.

CEILING
MOUNTED
SMOKE
DETECTOR

FIGURE 237. Permanently installed 120-volt unit.

The 120-volt type may be plugged into a convenient receptacle if desired. However, most people prefer not to have a cord running down the wall. Also, a plug-in unit may be accidentally disconnected.

To permanently connect the unit to the wiring system, proceed as follows:

1. *Locate the unit according to directions.*

2. *Mount the unit on wall or ceiling (Figure 237).*

3. *Connect to outlet box.*

   Install outlet box if necessary and run cable. Connect white to white and black to black wires with wire nuts. Connect the bare wire to the box with screw or clip.

4. *Test for operation as directed by manufacturer's instructions.*

IV. Installing Service Entrance Equipment

Service entrance equipment is that equipment necessary to supply a structure or building with power, provide overcurrent protection and a means of disconnect from its source. It is the first means of disconnect with overcurrent protection. See Code Section 100 for a more defined definition.

Many jurisdictions are now enforcing the strict rules found in the NEC concerning location of the service entrance equipment. Section 230-70 (a) requires that the service disconnecting means be installed at a readily accessible location either outside of the structure or inside near the point of entrance. This requirement has caused much frustration in the electrical industry because of a lack of understanding of the issues involved. A brief discussion should help eliminate this misunderstanding.

The conductors that connect the service utility to the customer's premises wiring is called a **service drop** or **lateral** (Figure 238). These conductors usually connect to the customer's service entrance conductors at the customer's weatherhead. These conductors are supplied from a distribution transformer and have no overload protection at the transformer output. Therefore, there is an unlimited bus in the event of a fault occurring in the customer's service entrance conductors up to the input of the service entrance panel board. At this point (service entrance equipment) overload protection, overcurrent protection and ground fault protection are provided.

Many buildings have burned to the ground because a fault occurred in the customer's service entrance conductors prior to reaching the service entrance equip-

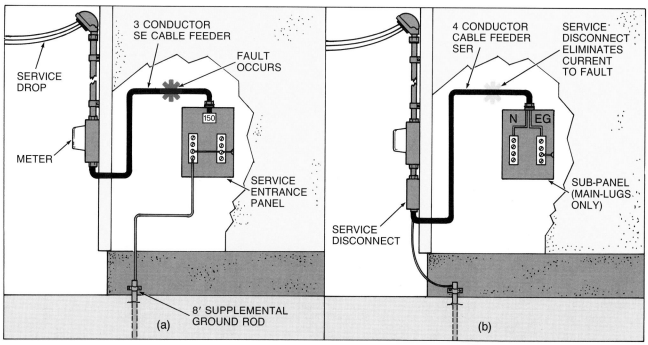

FIGURE 238. The service drop supplies power to the customer's property. (a) service equipment may consist of a meter and service entrance panel, or (b) a meter, service disconnect and sub-panel.

ment. When such a fault occurs, the serving utility will continue to supply unlimited current until there is nothing left to burn. This situation is particularly true where the service entrance equipment is located inside the structure (Figure 238a). For a number of years it was common practice to place the service entrance equipment in the middle of a structure such as a hallway. This placement worked well for the electrical contractor and the homeowner until a fault occurred. It is not uncommon for a service conductor to fault years after installation because a nail was driven into the conductor's installation during construction or for a variety of other conditions that cause the conductor to fail. The results are the same—destruction of the customer's equipment and property.

Code Section 230-70 provides guidelines for service entrance equipment locations that will eliminate this problem. This provision is not new although many authorities having jurisdiction have only recently begun enforcing the requirement.

Placing the service entrance equipment outside generally makes it inconvenient or uneconomical to run all the branch circuits from this location, although the Code will allow branch circuits to originate from this location. There is another solution to alleviate this problem.

Many utilities are now allowing their customers to purchase and install a self-contained meter base with the service disconnecting means as a part of the assembly (Figure 238A). When using this disconnect on the outside of a structure, the customer may elect to install a sub-panel in the interior of the structure (Figure 238b). The limitations as to how far into the structure this panel may be located center around the location of the major dedicated branch circuit. Other limitations may be dictated by other Code requirements.

Service entrance equipment has been discussed briefly and in general terms in previous sections. However, to install the equipment, much more information is required.

Upon completion of this section, you will be able to **select the type of service entrance equipment required and determine the size needed.** You will be able to **install the parts of the service entrance equipment and connect the necessary wiring.** Installing service entrance equipment is discussed as follows:

A. Determining Type and Size of Service Entrance Panel and Equipment.
B. Installing the Service Entrance Panel Cabinet.
C. Locating and Mounting the Service Drop Eyebolt or Insulator Rack.
D. Installing Service Entrance Cable or Conduit and Attachments.
E. Connecting Service Entrance Wiring.
F. Installing Ground Fault Circuit Interruptors (GFCI).
G. Connecting Circuits in Service Entrance Panel.

FIGURE 238A. Service entrance panel equipped with circuit breakers.

A. Determining Type and Size of Service Entrance Panel and Equipment

Determining types and sizes of SEP and equipment is discussed under the following headings:

1. Types of Service Entrance Panels and Equipment.
2. Determining Size of Service Entrance Panel and Equipment.

1. TYPES OF SERVICE ENTRANCE PANELS AND EQUIPMENT

Service entrance panels may be one of two types:
a. Circuit-Breaker-Type Service Entrance Panel.
b. Fuse-Type Service Entrance Panel.

a. Circuit-Breaker-Type Service Entrance Panel

More circuit-breaker-type SEP's than the fuse type are installed in new homes.

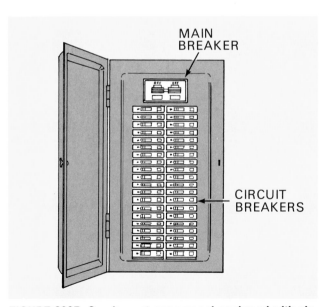

FIGURE 238B. Service entrance panel equipped with circuit breakers.

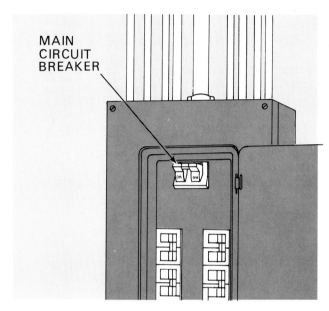

FIGURE 239. Main circuit breaker switches off all current to the service entrance panel.

Circuit-breaker entrance panels have several advantages over fuse types (Figure 238B). The switch on each breaker, when tripped, may be reset after the problem has been corrected without replacing it. Circuit breakers are easy to switch off when necessary to make repairs or for other reasons. Circuit breakers are available in sizes to protect any circuit desired. They are rated in amperes and will carry 80 percent of their rated loads continuously. They will take short periods of overload without tripping but provide protection from continued overloads or short circuits. Also, they cannot be tampered with and it is not easy to install a larger size accidentally or intentionally. The circuit breaker *main* is a heavy duty 2-pole on-off switch which can be used to turn off all power to the panel (Figure 239). It will trip when its rated current is exceeded.

b. Fuse-Type Service Entrance Panel

Fuse-type service entrance panels cost less to purchase but are less convenient to use (Figure 240). When a fuse "blows," it must be replaced unless equipped with special reset-type fuses.

Both fuse and circuit breaker SEPs are available with **main fuses** and **main breakers** that allow disconnecting of power to all circuits. The main fuses are mounted in a pull-out block (Figure 241).

For use with mobile homes, a combination of **meter base** and circuit breakers in a single cabinet is allowed in some areas (Figure 242).

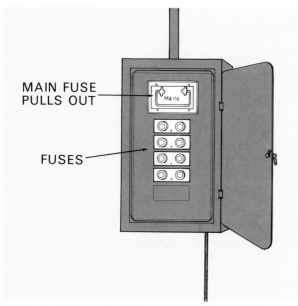

FIGURE 240. Fuse-type service entrance panel.

FIGURE 241. Main fuses are mounted in a pullout block.

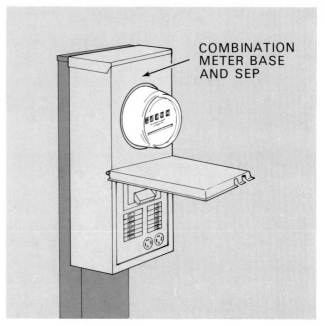

FIGURE 242. Combination meter base and circuit breaker service entrance panel in a single cabinet for mobile home installations.

---

2. DETERMINING SIZE OF SERVICE ENTRANCE PANEL AND EQUIPMENT

SEPs with 100, 150 or 200 amperes are commonly installed in homes. In larger homes, two panels are often used, for example, one 150-ampere and one 100-ampere.

> **NOTE: Article 220 of the code has been revised to replace the use of "watts" with "volt-amperes".**

The Code requires residential service entrance equipment to be rated for at least **100-ampere capacity unless (1) the computed load for the house is less than 10kVA (10000 VA)** or **(2) there are less than six 2-wire branch circuits in the house**. If either (1) or (2) applies, a 60-ampere service is allowed. Very few houses would qualify for either of the exceptions for 100-ampere service. For more information on size and rating, consult Code Section 230-42.

From a practical standpoint, you may consider 100 amperes as the minimum requirement for service entrance equipment rating in a home. If you install electric heat or air conditioning and an electric water heater, 150-ampere is the minimum and 200-ampere may be required. Power supplier representatives are usually available for advice on size of service entrance.

The Code provides guidelines for determining the minimum size of service entrance equipment for each individual home, including examples in Chapter 9 of the Code.

Demand factors are considered in computing the total load for a home. Use of the demand factor takes into account the difference between the maximum capacity of the system and the actual load it is likely to carry. **Appliances** have been tested and assigned a percentage demand factor to be used in computing the total load. Some **circuits** are also rated at less than their maximum capacity. For example, **each of the two small appliance circuits shown in the standard house plan as required by the code has a 2400 volt ampere capacity (120 volts × 20 amperes = 2400 volt-amperes). To convert to kVA, divide by 1000 = 2.4 kVA. The code allows them to be rated at 1500 volt-amperes (1.5 kVA) each because it is not likely that both will be fully loaded at the same time.**

To determine the size of service entrance required for the standard house plan by one method shown in the Code, proceed as follows:

The Code permits use of optional methods in computing the total load. Optional method from Code Section 220–30 may be used for single family dwellings with an ampacity of 100 amperes or greater. This would include most new homes.

Compute the size of service entrance required in the 215 square meter (2000 square foot) standard house plan.

Assume the 215 square meter (2000 square foot) house has the following:

—Two 20-ampere **small appliance** circuits.
—One 20-ampere **laundry** circuit.
—Two 5 kVA **wall mounted ovens**.
—One 5.8 kVA **counter-top cooking unit**.
—One 1.2 kVA **dishwasher**.
—One 4.5 kVA **water heater**.
—One 8.5 kVA **clothes dryer** (high speed).
—Two 1.2 kVA **built-in bathroom heaters**.
—One 9.6 kVA **central air conditioner**.

Follow the procedures from optional method **demand factors** in the Code and from Table IX.

CALCULATE "OTHER LOAD"

1. *Include 1500 volt-amperes for each two-wire, 20 ampere small appliance branch circuit.*

   Rated by the code at 1500 volt-amperes each; 2 × 1500 = 3000 or = 3 kVA.

2. *Three volt-amperes per 0.09 square meter (square foot) for general lighting and general-use receptacles.*

   Multiply 28 times the number of square meters (3 times the number of square feet) in the house, exclusive of open porches and garages.

   3 volt-amperes × 2000 square feet
   = 6000 or = 6 kVA
   Total steps 1 and 2 = 9 kVA

TABLE IX. OPTIONAL CALCULATION FOR DWELLING UNIT LOAD IN kVA
(From Code Table 220–30)

Largest of the following five selections.

(1) 100 percent of the nameplate rating(s) of the air conditioning and cooling, including heat pump compressors.

(2) 100 percent of the nameplate ratings of electric thermal storage and other heating systems where the usual load is expected to be continuous at the full nameplate value. Systems qualifying under this selection shall not be figured under any other selection in this table.

(3) 65 percent of the nameplate rating(s) of the central electric space heating including integral supplemental heating in heat pumps.

(4) 65 percent of the nameplate rating(s) of electric space heating if less than four separately controlled units.

(5) 40 percent of the nameplate rating(s) of electric space heating of four or more separately controlled units.

Plus: 100 percent of the first 10 kVA of all other load. 40 percent of the remainder of all other load.

3. *List the nameplate ampere rating of all fixed appliances, ranges, wall-mounted ovens and counter-mounted cooking units.*

Check the nameplates of each item included and list the volt-amperes of each.

Fixed Appliances	Volt Amperes
Laundry circuit (rated by Code at 1500 volt-amperes) or	1.5 kVA
Two wall-mounted ovens (5000 volt amperes each) 10,000 or	10.0 kVA
Countertop cooking unit 5800 or ...	5.8 kVA
Dishwashwer 1200 or	1.2 kVA
Water Heater 4500 or	4.5 kVA
Total "Other Load"	**23.0 kVA**
9 kVA plus 23 kVA = 32.0 kVA	

4. *List the nameplate ampere or kVA rating of all motors and of all low-power-factor loads.*

Include here only those motors and other loads not included in steps 3 and 5. If the motor rating is given in amperes, convert to volt-amperes by multiplying volts × amperes. For example, assume a motor rated 9 amperes, 240 volts: 240 × 9 = 2160 volt-amperes or 2.16 kVA.

NOTE: Code Section 220–21 states that where it is unlikely that two dissimilar loads will be in use at the same time, it is permissible to omit the smaller of the two in computing the total load. This applies to heating and air conditioning loads described in Step 5 as follows:

5. *Use the largest of the following:*

(a) 100 percent of the nameplate rating(s) of the air conditioning and cooling, including heat pump compressors.

(b) 65% of the nameplate rating(s) of the central electric space heating, including integral supplemental heating in heat pumps.

(c) 65% of the nameplate rating(s) of electric space heating if less than four separately controlled units.

(d) 40% of the nameplate rating(s) of electric space heating of four or more separately controlled units.

Computations for the standard house plan are as follows:

Air conditioning included in total at 9.6 kVA.

6. *Compute calculated load.*

Refer to Table IX.

100% of nameplate rating of the air conditioner
. . . . 9.6 kVA

100% of the first 10 kVA of all other load
. . . . 10.0 kVA

40% of the remainder of all other load
(22 kVA × 40%) . . . . 8.8 kVA

Total calculated load 28.4 kVA

28.4 kVA = 28400 VA ÷ by 240 volts = 117 ampere service rating. Use 150-ampere service rating.

Allowances should always be made for some future expansion of the load.

This home may be served by 150-ampere service, which provides enough capacity for future needs. The service entrance cable or wires, the main disconnect switch and the service entrance panel must be sized for 150 amperes.

TABLE X. CONDUCTOR TYPES AND SIZES IN COPPER AND ALUMINUM FOR SERVICE AND FEEDER CONDUCTORS
(Notes to Ampacity Tables 310–16 thru 310–19, note 3)

Reprinted with permission from NFPA 70–1993, the National Electrical Code®, Copyright 1992, National Fire Protection Association, Quincy, MA 02269. This reprinted material is not the complete and official position of the National Fire Protection Association, on the referenced subject which is represented only by the standard in its entirety. Conductor Types and Sizes RH-RHH-RHW-THHW-THW-THWN-THHN-XHHW-USE

Copper	Aluminum and Copper-Clad AL	Service Rating in Amperes
AWG	AWG	
4	2	100
3	1	110
2	1/0	125
1	2/0	150
1/0	3/0	175
2/0	4/0	200
3/0	250 kcmil	225
4/0	300 kcmil	250
250 kcmil	350 kcmil	300
350 kcmil	500kcmil	350
400 kcmil	600 kcmil	400

Some local codes specify minimum wire sizes for a service entrance of a given capacity. For example, No. 2 THW wire with a 100-ampere service, No. 1/0 for 150-ampere and No. 3/0 for 200-ampere.

The Code in "Notes to Tables 310–16 through 310–19" provides a simple method to determine required service entrance conductor and feeder conductor sizes for residences, as follows:

Conductors for a three-wire single phase dwelling unit may be sized according to Table X. The grounded service entrance conductors may be two AWG sizes smaller than the ungrounded conductors (hot wires) provided the requirements of Section 230-42 are met. They will usually be met when two or more 240-volt loads are included in the calculations, for example a water heater and clothes dryer.

B. Installing the Service Entrance Panel Cabinet

The **cabinet** is sometimes called the box (Figure 243). It is the steel or nonmetallic cabinet with hinged door in which are mounted the service entrance terminals, circuit terminals and circuit breakers or fuses.

To mount the SEP, proceed as follows:

1. *Remove cover.*

   Remove cover mounting screws (usually four), and replace cover in carton for protection.

2. *Mount the cabinet.*

   For **flush mount**, line up front edge of cabinet making allowance for the finished wall surface. Screw (preferred) or nail to stud through small knockouts or holes on each side. For **surface mount**, screw or nail through keyhole slots in back of box.

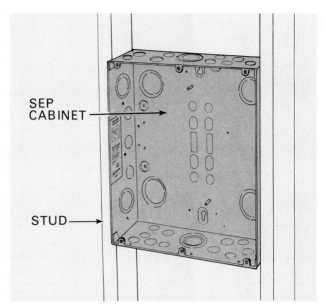

SEP CABINET

STUD

FIGURE 243. Service entrance panel cabinet encloses the service entrance terminals, circuit terminals and circuit breakers or fuses.

C. Locating and Mounting the Service Drop

Locating and mounting the service drop eyebolt or insulator rack are discussed under the following headings:

1. Locating the Service Drop Insulator or Insulator Rack.
2. Mounting the Service Drop Insulator or Insulator Rack.

1. LOCATING THE SERVICE DROP INSULATOR OR INSULATOR RACK

Before you install the service entrance equipment, you must know the location of the **service drop** (wires from the power source to the house). Location of the service drop depends to some extent on the location of the service entrance panel. Instructions for locating the service entrance panel were given in Section II, "Installing Wiring."

The **service drop** is usually located and installed by the power supplier. It includes the service conductors required to reach from the street main, transformer or meter pole, to the house (Figure 244). Some power suppliers furnish and install the insulator, or **insulator rack**, required for the service drop. In most areas, it is installed by the electrician. Locate the service drop attachment at the location agreed upon with the power supplier.

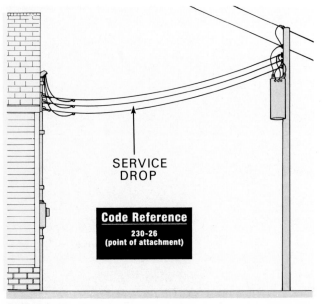

SERVICE
DROP

Code Reference
230-26
(point of attachment)

FIGURE 244. The service drop is the cable or wires that extend from the power company transformer to the house or building. It is usually located and installed by the power supplier.

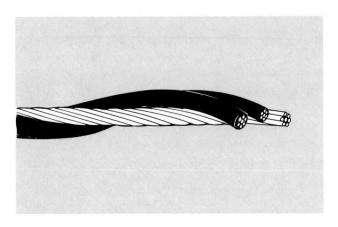

FIGURE 245. Triplex cable, made up of a bare neutral wire with two insulated cables, is used for most service drops to new homes.

The service drop conductors for a 120/240-volt installation may be three separate wires including two black insulated wires and a neutral wire (Figure 244). The neutral may be a white insulated wire but usually it is bare.

Most service drop installations to new homes are made by use of triplex cable, which is made up of a bare neutral wire with the two insulated cables wrapped around it for support (Figure 245). Proceed as follows:

1. *Select tentative location for insulator or insulator rack.*

2. *Check ground clearance (Figure 246).*

   The Code requires that service drop conductors, where not in excess of 600 volts, nominal, shall have the following minimum clearance from final grade.

   3.05m (10 feet)—at the electric service entrance to buildings or at the drip loop of the building electric entrance or above areas or sidewalks accessible only to pedestrians measured from final grade or other accessible surface only for service drop cables supported on and cabled together with a grounded bare messenger and limited to 150 volts to ground.

   3.66 m (12 feet)—for those areas listed in the 4.57 m (15 feet) classification when the voltage is limited to 300 volts to ground.

   4.57 m (15 feet)—over residential property and driveways, and those commercial areas not subject to truck traffic.

   5.49 m (18 feet)—over public streets, alleys, roads, parking areas subject to truck traffic, driveways on other than residential property, and other land traversed by vehicles such as cultivated, grazing, forest, and orchard.

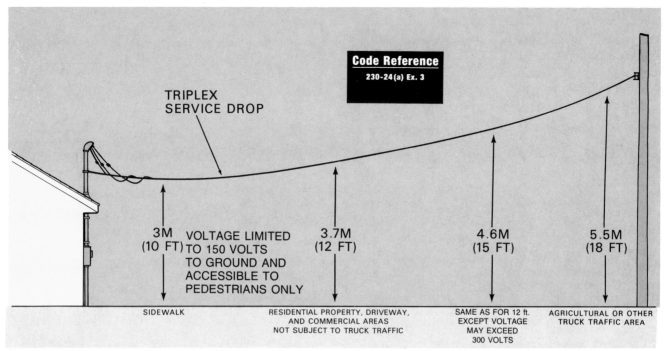

TRIPLEX SERVICE DROP

Code Reference
230-24(a) Ex. 3

3M (10 FT) — VOLTAGE LIMITED TO 150 VOLTS TO GROUND AND ACCESSIBLE TO PEDESTRIANS ONLY

3.7M (12 FT)

4.6M (15 FT)

5.5M (18 FT)

SIDEWALK

RESIDENTIAL PROPERTY, DRIVEWAY, AND COMMERCIAL AREAS NOT SUBJECT TO TRUCK TRAFFIC

SAME AS FOR 12 ft. EXCEPT VOLTAGE MAY EXCEED 300 VOLTS

AGRICULTURAL OR OTHER TRUCK TRAFFIC AREA

FIGURE 246. Service drop conductors have minimum clearance requirements, depending on the sidewalks, or traffic areas to be crossed.

3. *Check clearance over roof.*

If the service drop conductors pass over a roof, the minimum clearance above the roof must be as follows:

If the roof pitch is **not less than 4 in 12 inches**, and voltage between conductors is not over 300, a 0.9 m (3 feet) clearance is permitted.

—**0.46 meters (18 inches) if terminated at a mast** and not more than 1.83 meters (6 feet) of the overhang portion of the roof is crossed by the service drop (Figure 247A).

If passing over the roof of a building, conductors must have a vertical clearance of at least 2.44 meters (8 feet) from the roof surface. Horizontal clearance must be at least 914 millimeters (3 feet).

4. *Check clearance for windows and other items.*

Code Section 230-9 says conductors shall have a clearance of not less than 0.9 meters (3 feet) from windows, doors, porches, fire escapes, or similar locations.

Conductors run above the top level of a window shall be considered out of reach from that window.

5. *Check power supplier requirements.*

Some power suppliers have requirements that are different from the Code for service drop mounting. Check with your power supplier for instructions.

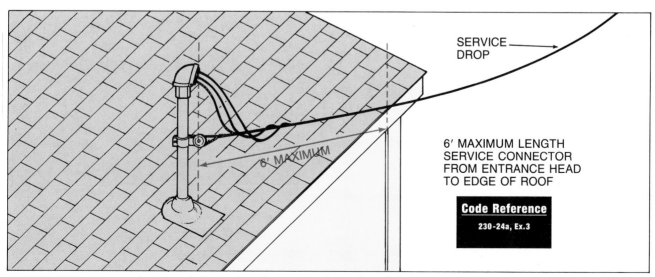

SERVICE DROP

6' MAXIMUM

6' MAXIMUM LENGTH SERVICE CONNECTOR FROM ENTRANCE HEAD TO EDGE OF ROOF

Code Reference
230-24a, Ex.3

FIGURE 247A. The service drop is limited to six feet maximum length from the entrance head to the roof edge.

121

2. MOUNTING THE SERVICE DROP INSULATOR OR INSULATOR RACK

Proceed as follows:

1. *At the selected location, fasten the cable to the side of the house by means of an insulator rack or to a mast through the roof, usually made of 5.1 cm (2 inch) conduit (Figure 248).*

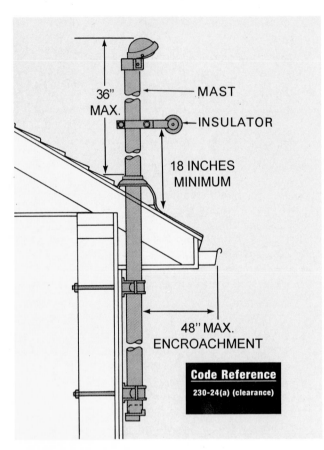

FIGURE 248. Two-inch conduit mast through the roof.

2. *Locate the entrance head above the service drop to allow room for the drip loops (Figure 248A).*

The drip loops prevent water from entering the entrance head. Proceed as follows:

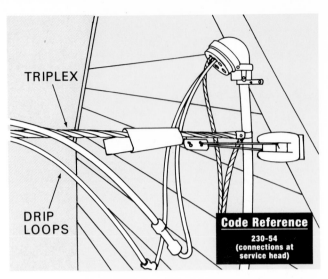

FIGURE 248A. The entrance head must be mounted higher than the service drop to allow room for drip loops.

If mounting a screw-type insulator, screw it directly into a stud, plate or other reinforcement in back of the siding (Figure 248B).

If mounting an insulator rack, fasten the insulator rack with lag screws.

Both types must be fastened securely to withstand the strain of the service drop during extreme weather conditions.

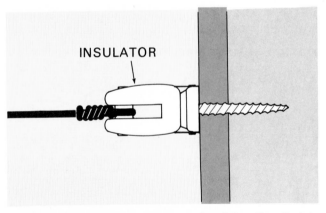

FIGURE 248B. Fasten screw-type insulator directly into a stud or other framing member.

122

D. Installing Service Entrance Cable or Conduit and Attachments

Your local code may determine the type of installation you will use for the service entrance wiring and attachments. Four methods may be used for the installation. They are discussed as follows:

1. Installing Service Entrance Cable and Meter Base.
2. Installing Service Entrance Conductors in Conduit.
3. Installing a Mast Type of Service Entrance.
4. Installing Underground Service Entrance Cable.

1. INSTALLING SERVICE ENTRANCE CABLE AND METER BASE

Wire used for service entrance cable may be:
a. Aluminum.
b. Copper.

a. Aluminum

Aluminum wire or cable is often installed for the service entrance because it costs less than copper. The following procedures are recommended when connecting aluminum wire:

1. *Be sure the service entrance terminals are labeled Al-Cu.*
2. *Strip the insulation, being careful not to nick the wires.*
3. *Wire-brush the conductor strands.*
4. *Thoroughly coat the stripped conductor with a suitable anti-oxidant compound such as ALNOX or PENETROX A13.*
5. *Insert conductor and tighten connector screw very tightly.*

b. Copper

Copper is very satisfactory for service entrance installation but in larger sizes, it is expensive. To install copper, use care when stripping insulation so you don't nick the wires. Also, be sure to tighten the connector screw very tightly. **IMPORTANT:** Most connectors will have detailed instructions concerning torque requirements.

Service entrance cable is used for this installation.

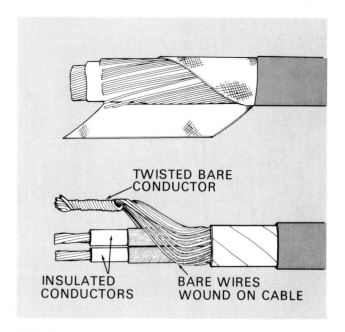

TWISTED BARE CONDUCTOR

INSULATED CONDUCTORS   **BARE WIRES WOUND ON CABLE**

FIGURE 249. Service entrance cable with strands of bare neutral wire wound around the insulated conductors. Unwrap the separate strands and twist to form the neutral wire.

The bare neutral wire of the cable is made up of many small wires that are wound around the insulated wires (Figure 249). Proceed as follows to install the service entrance cable and meter base:

1. *Measure and cut cable for installation between entrance head and meter base.*
   (1) *Determine from power supplier or other authority the locations of the entrance head and meter base.*
   (2) *Mark the location of each.*
   (3) *Measure the distance between the two marks and add 1.1 meter (3½ feet) for connections at the entrance head and meter base.*
   (4) *Cut the measured length of cable.*

2. *Remove the outer jacket from cable.*

   The outer layer of heavy plastic jacket and an inner layer of lighter material are both removed for a length of about 0.9 m (3 feet).

3. *Prepare neutral wire for installation.*

   Unwrap the separate strands, pull to one side and twist together to form a bare conductor (Figure 249).

   NOTE: If the neutral wire is insulated, it must be unmistakably identified at each end.

123

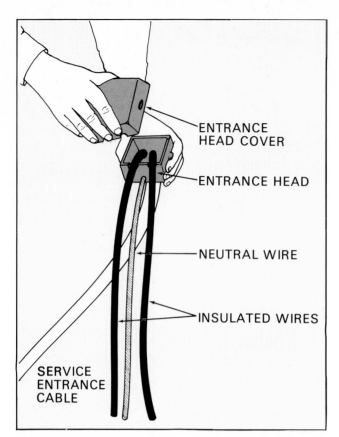

ENTRANCE
HEAD COVER

ENTRANCE HEAD

NEUTRAL WIRE

INSULATED WIRES

SERVICE
ENTRANCE
CABLE

FIGURE 250. Insulated wires are inserted through the holes provided in the entrance head and the neutral wire through the notch or other type opening.

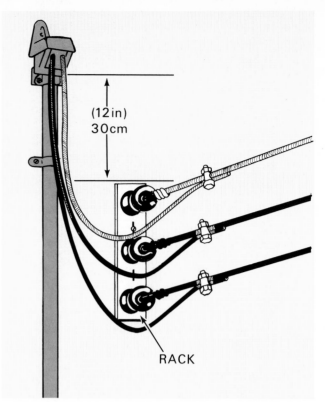

(12 in)
30 cm

RACK

FIGURE 251. Entrance head should be 30 cm (12 inches) above the rack. Each connection must be wrapped with tape for insulation.

4. *Install wires in entrance head (Figure 250).*

Remove screws from entrance head (weather-head) to disassemble. Insert the insulated wires through the bushing holes provided. Insert the neutral wire through the notch below the bushing and assemble the entrance head. Cable covering must be under the clamp of the entrance head to prevent damage to conductors. Do not clamp too tight; insulation may be damaged. You should have about one meter (three feet) of wire extended from the entrance head. These wires are later connected to the service drop wires by the **power supplier** when the installation is complete.

5. *Install the entrance head.*

Mount the entrance head 30 cm (12 inches) higher than the rack and in line with the meter location (Figure 251). Fasten in place with screw(s).

6. *Install the meter base.*

Check with the power supplier for recommendations on meter height.

If none, mount the meter base 1.4 to 1.7 m (4 ½ to 5 feet) above ground at the marked location. Fasten screws in holes in the back. The meter is later installed by the power company (Figure 252).

7. *Attach connectors to meter base (Figure 253).*

Install watertight connector at top and regular connector at bottom outlets of the meter base. Do not tighten the part holding the cable.

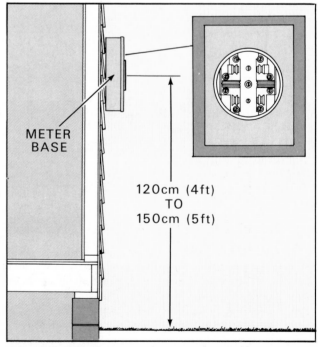

METER
BASE

120 cm (4 ft)
TO
150 cm (5 ft)

FIGURE 252. Locate meter base at height of 1.4 to 1.7 m (4 ½ to 5 feet) or as recommended by your power supplier. Fasten screws in holes provided in the back.

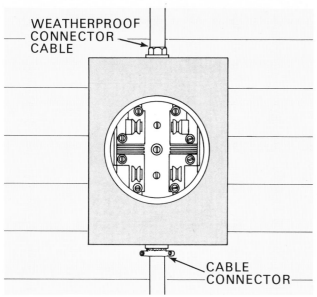

FIGURE 253. Watertight connector installed at upper opening in the meter base.

8. *Insert the end of cable from entrance head through the top meter base opening and tighten connector on the cable (Figure 254).*

   You should have about 15 to 20 cm (6 to 8 inches) of cable inside the meter base for terminal connections.

9. *Drill hole through wall for cable.*

   Locate the hole for the shortest or most convenient route to reach the SEP inside the house.

10. *Measure and cut cable length to reach service entrance panel.*

    Add 35 to 60 cm (14—24 inches) to the measured distance for meter connections and SEP connections.

11. *Install cable between meter base and SEP.*

    (1) *Insert wires in bottom outlet of meter base.*

       Strip cable covering from insulated conductors.

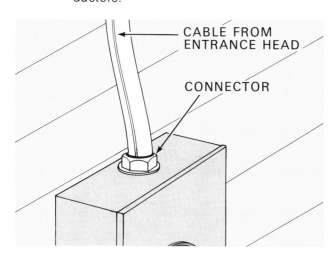

FIGURE 254. Cable from entrance head inserted through top of meter base.

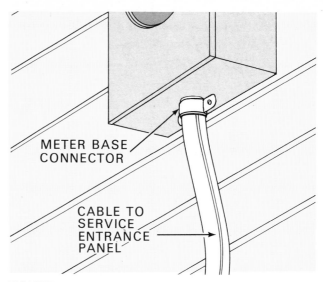

FIGURE 254. Cable from entrance head inserted through top of meter base.

(2) *Insert cable into meter base connector and tighten (Figure 255).*

(3) *Insert cable through the wall and pull to SEP (Figure 256).*

   Strip outside cable covering from wires that go into the SEP.

   Remove knockout, install connector, and insert cable. Approximately 15 to 35 cm (6 to 14 inches) of cable should extend inside the panel box for terminal connections. Exact length depends on location of knockout opening and position of lugs in the panel.

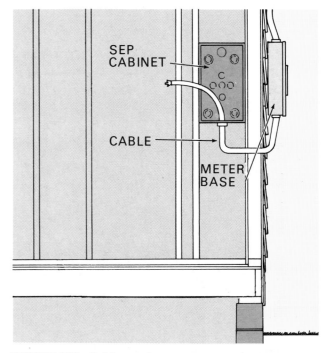

FIGURE 256. Cable run from meter base to service entrance panel.

125

If cable goes through a masonry wall, do not concrete around the cable. Use a mastic-type sealer.

(4) Install sill plate (if required) and sealer (Figure 257).

Many localities do not require a sill plate.

At the point where the cable enters the building, install the sill plate where required to seal the hole. Apply a sealing compound to any space between the cable and sill plate, and where cable enters the building if no sill plate is used.

(5) Install cable straps (Figure 258).

Service entrance cable must be supported by straps or other means within 30 cm (12 inches) of every service head, gooseneck, or connection and at least every 76.5 cm (30 inches) in between. Install straps within 30 cm (12 inches) of the meter connections and SEP.

12. Connect cable to meter base terminals (Figure 259).

(1) Strip cable covering from the conductors and strip the insulation from the ends of the two insulated conductors.

(2) Apply inhibitor paste and connect one insulated conductor to an upper (line) terminal and the second to the other upper (line) terminal.

(3) Connect the bare neutral conductor to the center neutral terminal.

Some power suppliers prefer not to have the neutral conductor cut for this connection.

(4) Repeat steps 1–3 for conductor connections to lower (output) terminals and to the neutral terminal.

(5) Force the conductors against the back of the meter base.

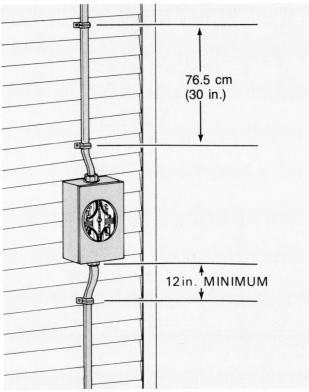

FIGURE 258. Service entrance cable supported by straps or other support every 76.5 cm (30 inches) and within 30 cm (12 inches) of every entrance head, gooseneck or other connection.

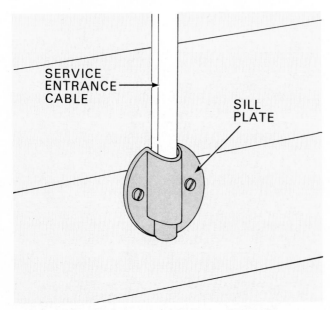

FIGURE 257. Sill plate installed in wall opening.

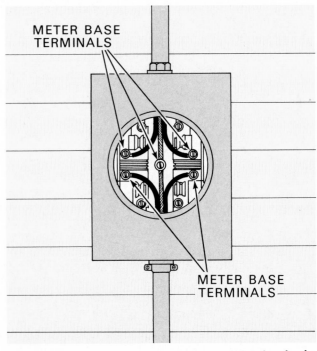

FIGURE 259. Meter base terminals completely wired.

2. INSTALLING SERVICE ENTRANCE CONDUCTORS IN CONDUIT

Some details of conduit installations are given in Section V., "Installing Metallic Conduit." However, you can assemble the required conduit and connections for rigid conduit, IMC (intermediate metal conduit) or EMT (electrical metallic tubing) service entrance requirements using wrenches, a hacksaw with 13 teeth per centimeter (32 teeth per inch), a pipe reamer and threaders for the rigid conduit. The Code now requires the use of **standard conduit cutting dies** with a 0.63 mm taper per centimeter (¾-inch taper per foot) for threading rigid conduit. EMT conduit cannot be threaded on the job because it is very thin-walled. Use connectors and couplings for EMT.

Other than assembling the conduit, procedures for conduit service entrance installation are similar to cable, but there are some notable differences.

First, there is a difference in the wire used inside the conduit. Every wiring installation you have learned to this point has used cable with 2 or more wires. **In conduit wiring, separate insulated wires are pulled into the conduit (Figure 260).** The conduit provides protection against weather and mechanical damage so outer cable covering is not required. Other differences in installation are pointed out in procedures.

Wires used inside service entrance conduit may be Type TW, THW, RHW or other types approved for that use. The neutral wire may be bare, but if insulated it should be white. Colored or white wires may not be available in every size. If not, you must identify the neutral wire by painting it white, wrapping with tape or with a tag at each terminal point. The other two wires may be black or one red and one black.

Installing service entrance wiring in conduit is discussed as follows:

> NOTE: The Code now permits the use of **intermediate metal conduit (IMC) wherever rigid conduit is required in residential wiring. The wall thickness of IMC is thinner than rigid conduit but thicker than EMT. Procedures for installing IMC are the same as for rigid conduit.**

a. Installing Service Entrance Wiring in Rigid Conduit.
b. Installing Service Entrance Wiring in Thinwall Conduit (EMT).

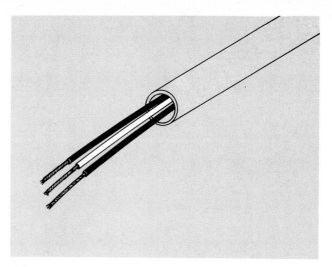

FIGURE 260. Separate insulated wires are installed in conduit.

a. Installing Service Entrance Wiring in Rigid Conduit

Rigid metal conduit is steel pipe that looks very much like water pipe (Figure 261). But water pipe may never be substituted for conduit. Rigid metal conduit has a rust resistant finish inside and out. The inside finish is smooth to prevent damage to wires as they are pulled through, which is not the case with water pipe. Metal conduit must carry the underwriters label to meet Code requirements. It is soft enough to bend as required.

For service entrance installations and other small jobs, you may use fittings instead of bending the conduit.

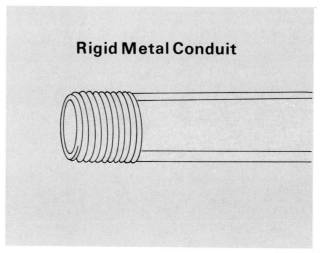

Rigid Metal Conduit

FIGURE 261. Rigid metal conduit looks much like common water pipe but there are important differences.

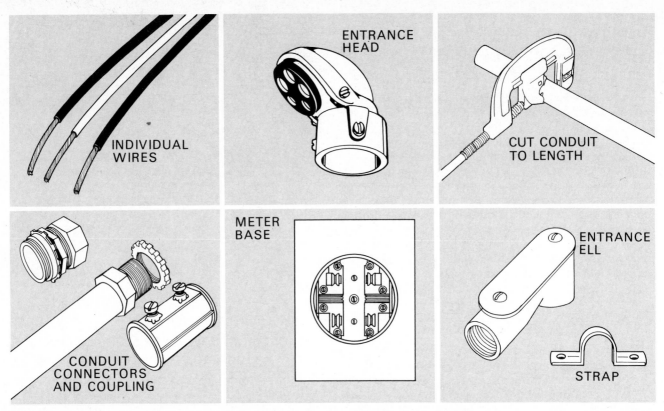

FIGURE 262. Materials required to install service entrance wiring in rigid conduit.

The materials required for installing 120/240-volt entrance wiring in rigid conduit include (Figure 262) three individual wires, entrance head, conduit cut to measured length, conduit connectors, meter base, entrance ell and straps. The entrance **head** must be the type made for conduit (Figure 263). The entrance **ell** is installed where the conduit enters the house. It has a removable back to make it easier to install the wires.

The conduit is assembled first and wires are pulled through after the assembly is mounted on the wall. Proceed as follows:

1. *Determine size conduit to install.*

   Table XI shows maximum number of wires that can be installed in conduit of given sizes. Select conduit of correct size for your service entrance.

FIGURE 263. Entrance head with fitting for rigid conduit installation.

128

TABLE XI. MAXIMUM NUMBER OF CONDUCTORS IN TRADE SIZES OF CONDUIT OR TUBING
(Code Tables 3A, 3B, and 3C)
(Based on Code Table 1, Chapter 9) - Pages 821, 822, 823

Type Letters	Conductor Size AWG, kcmil	½	¾	1	1¼	1½	2	2½	3	3½	4	5	6
TW, XHHW (14 thru 18) RH (14 + 12)	14	9	15	25	44	60	99	142	171				
	12	7	12	19	35	47	78	111	131	176			
	10	5	9	15	26	36	60	85	131	176			
	8	2	4	7	12	17	28	40	62	84	108		
RHW and RHH (without outer covering), RH (10 + 8), THW, THHW	14	6	10	16	29	40	65	93	143	192			
	12	4	8	13	24	32	53	76	117	157			
	10	4	6	11	19	26	43	61	95	127	163		
	8	1	3	5	10	13	22	32	49	66	85	133	
TW,	6	1	2	4	7	10	16	23	36	48	62	97	141
	4	1	1	3	5	7	12	17	27	36	47	73	106
THW,	3	1	1	2	4	6	10	15	23	31	40	63	91
	2	1	1	2	4	5	9	13	20	27	34	54	78
	1		1	1	3	4	6	9	14	19	25	39	57
FEPB (6 thru 2), RHW and RHH (without outer covering	1/0		1	1	2	3	5	8	12	16	21	33	49
	2/0		1	1	1	3	5	7	10	14	18	29	41
	3/0		1	1	1	2	4	6	9	12	15	24	35
	4/0			1	1	1	3	5	7	10	13	20	29
	250			1	1	1	2	4	6	8	10	16	23
	300			1	1	1	2	3	5	7	9	14	20
	350				1	1	1	3	4	6	8	12	18
	400				1	1	1	2	4	5	7	11	16
RH, THHW	500					1	1	1	3	4	6	9	14
	600						1	1	3	4	5	7	11
	700						1	1	2	3	4	7	10
	750						1	1	2	3	4	6	9
THWN,	14	13	24	39	69	94	154						
	12	10	18	29	51	70	114	164					
	10	6	11	18	32	44	73	104	160				
	8	3	5	9	16	22	36	51	79	106	136		
THHN, FEP (14 thru 2) FEPB (14 thru 8), PFA (14 thru 4/0) PFAH (14 thru 4/0) Z (14 thru 4/0) XHHW (4 thru 500 kcmil)	6	1	4	6	11	15	26	37	57	76	98	154	
	4	1	2	4	7	9	16	22	35	47	60	94	137
	3	1	1	3	6	8	13	19	29	39	51	80	116
	2	1	1	3	5	7	11	16	25	33	43	67	97
	1		1	1	3	5	8	12	18	25	32	50	72
	1/0		1	1	3	4	7	10	15	21	27	42	61
	2/0		1	1	2	3	6	8	13	17	22	35	51
	3/0		1	1	1	3	5	7	11	14	18	29	42
	4/0		1	1	1	2	4	6	9	12	15	24	35
	250			1	1	1	3	4	7	10	12	20	28
	300			1	1	1	3	4	6	8	11	17	24
	350			1	1	1	2	3	5	7	9	15	21
	400				1	1	1	3	5	6	8	13	19
	500					1	1	2	4	5	7	11	16
	600					1	1	1	3	4	5	9	13
	700						1	1	3	4	5	8	11
	750						1	1	2	3	4	7	11
XHHW	6	1	3	5	9	13	21	30	47	63	81	128	185
	600					1	1	1	3	4	5	9	13
	700						1	1	3	4	5	8	11
	750						1	1	2	3	4	7	10

Note 1: This table is for concentric stranded conductors only. For cables with compact conductors, the dimensions in Table 5A shall be used. Note 2: Conduit fill for conductors with a -2 suffix is the same as for those types without the suffix.

Table XI continued on next page

Table XI continued from previous page

TABLE XI. MAXIMUM NUMBER OF CONDUCTORS IN TRADE SIZES OF CONDUIT OR TUBING
(Code Tables 3A, 3B, and 3C)
(Based on Code Table 1, Chapter 9) - Pages 821, 822, 823

Conduit or Tubing Trade Size (Inches)		½	¾	1	1¼	1½	2	2½	3	3½	4	5	6
Type Letters	Conductor Size AWG, kcmil												
RHW,	14	3	6	10	18	25	41	58	90	121	155		
	12	3	5	9	15	21	35	50	77	103	132		
	10	2	4	7	13	18	29	41	64	86	110		
	8	1	2	4	7	9	16	22	35	47	60	94	137
RHH	6	1	1	2	5	6	11	15	24	32	41	64	93
	4	1	1	1	3	5	8	12	18	24	31	50	72
(with	3	1	1	1	3	4	7	10	16	22	28	44	63
outer	2		1	1	3	4	6	9	14	19	24	38	56
covering)	1		1	1	1	3	5	7	11	14	18	29	42
	1/0		1	1	1	2	4	6	9	12	16	25	37
	2/0			1	1	1	3	5	8	11	14	22	32
	3/0			1	1	1	3	4	7	9	12	19	28
	4/0			1	1	1	2	4	6	8	10	16	24
	250				1	1	1	3	5	6	8	13	19
	300				1	1	1	3	4	5	7	11	17
	350				1	1	1	2	4	5	6	10	15
	400					1	1	1	3	4	6	9	14
	500				1	1	1	1	3	4	5	8	11
	600					1	1	1	2	3	4	6	9
	700					1	1	1	1	3	3	6	8
	750						1	1	1	3	3	5	8

Note 1: This table is for concentric stranded conductors only. For cables with compact conductors, the dimensions in Table 5A shall be used.
Note 2: Conduit fill for conductors with a -2 suffix is the same as for those types without the suffix.

The number of conductors allowed in Table XI for various sizes of conduit applies to all conduit wiring installations. They are not intended to apply to short sections of conduit used to protect exposed wiring from physical damage according to the Code. A general recommendation is to install 3.8 cm or 5 cm (1½ or 2 inch) conduit for the house service entrance.

2. *Install service-drop insulator rack as for cable.*

3. *Locate entrance head as for cable.*

4. *Measure space between locations of service entrance head and meter, between meter and entrance ell and from entrance ell to service entrance cabinet.*

5. *Cut measured lengths of conduit with a pipe cutter or hacksaw (Figure 264). File or ream ends of each piece to remove rough edges (Figure 265).*

This *must* be done to prevent damage to insulation as wires are pulled through.

6. *Thread conduit where required with tapered conduit dies.*

The top end of the entrance head may be threaded or not depending on the type of entrance head used.

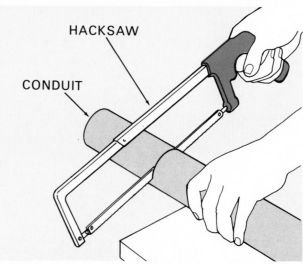

HACKSAW

CONDUIT

FIGURE 264. Cut rigid conduit with pipe cutter or fine-tooth hacksaw.

7. *Assemble parts on the ground.*

(1) *Attach entrance head.*

If the entrance head is threaded, screw it on the threaded end of the conduit (Figure 266). If not threaded, place on end of con-

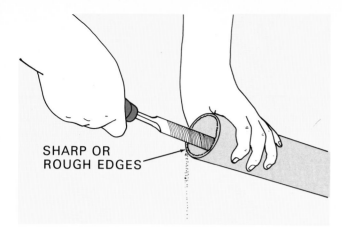

FIGURE 265. File or ream ends of the conduit to remove sharp or rough edges that might cut wiring insulation.

SHARP OR ROUGH EDGES

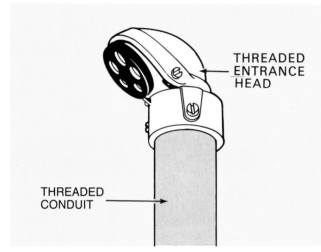

THREADED ENTRANCE HEAD

THREADED CONDUIT

FIGURE 266. If entrance head is threaded, screw it on the conduit end.

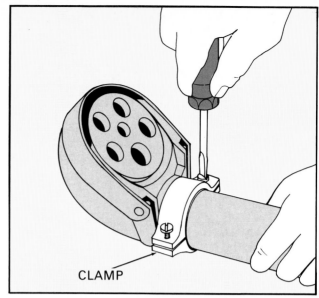

CLAMP

FIGURE 267. For bolt-on entrance heads, tighten the bolts that clamp the conduit.

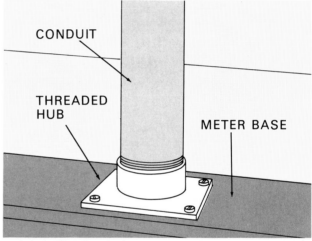

CONDUIT

THREADED HUB

METER BASE

FIGURE 268. Attach hub to meter base and screw conduit into the threaded hub.

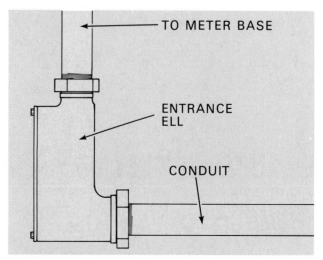

TO METER BASE

ENTRANCE ELL

CONDUIT

FIGURE 269. Install entrance ell and connect short length of conduit to reach service entrance cabinet.

duit and tighten bolts or set screw(s) on each side of clamp (Figure 267).

(2) *Connect conduit to meter (Figure 268).*

Most power suppliers require a threaded hub on top of the meter base to make the connection water tight. Attach the hub to the meter base, usually with four bolts. Insert threaded conduit and tighten. If conduit enters the meter base at bottom, or side, use two locknuts, one inside and one outside. Attach locknut to threaded conduit end. Screw the locknut on as far as it will go. Place conduit through opening in meter base, attach inside locknut and turn until snug. Turn outside locknut until very tight.

(3) *Connect entrance ell.*

Attach entrance ell to threaded conduit length from meter base (Figure 269).

(4) *Connect short length of conduit between entrance ell and service entrance cabinet.*

This completes the assembly of all outside connections (Figure 269).

131

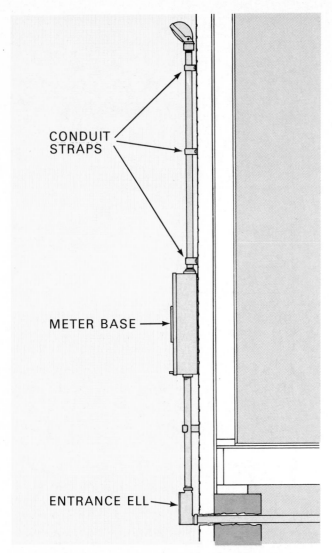

CONDUIT STRAPS

METER BASE →

ENTRANCE ELL →

FIGURE 270. Install the assembly on the wall and fasten in place with straps.

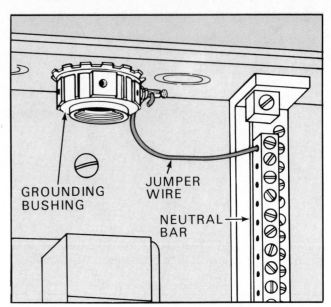

GROUNDING BUSHING

JUMPER WIRE

NEUTRAL BAR →

FIGURE 271. Jumper wire connects the grounding bushing and neutral bar.

8. *Install the assembly on the wall.*

Insert the short length of conduit through the wall with the assembly held upright. Fasten the conduit to the building with conduit straps (Figure 270). Install support brackets behind the conduit where mounted away from the wall as in Figure 270.

It must be supported within 0.9 m (3 feet) of every connection, plus additional supports not over 3.1 m (10 feet) apart for 1.3 and 1.9 cm (½ and ¾ inches), not over 3.7 m (12 feet) for 2.5 cm (1 inch), 4.3 m (14 feet) for 3.2 and 3.8 cm (1 ¼ and 1 ½ inches) and 4.9 m (16 feet) for 5.1 and 6.4 cm (2 and 2 ½ inches).

9. *Connect conduit to service entrance cabinet.*

Attach locknut and grounding bushing. Twist locknut tight to insure solid grounding connection. Run a jumper wire from one bushing to the neutral bar (Figure 271). (Refer to Code Table 250 – 94 for correct size conductor.)

10. *Install the wires.*

Measure and cut the first set of wires between entrance head and meter base so you will have 0.9 m (36 inches) extending from the entrance head and 20 to 25 cm (8 to 10 inches) inside the meter base.

Remove front cover from meter base and push wires up through meter opening through conduit to entrance head. Remove top of entrance head and thread wires through bushing holes. Replace top on entrance head and leave 0.9 m (3 feet) (or length recommended by your power supplier) of wire extended for connection to service drop (Figure 270).

Measure and cut second set of wires to reach from bottom terminals of meter base to terminals inside the SEP. Push wires through bottom opening of meter base toward the entrance ell. Remove cover on entrance ell to push wires through more easily. Pull wires into service entrance cabinet. Replace cover on entrance ell (Figure 270).

11. *Connect wiring to meter base.*

Follow procedures in Step 12 Under IV. D. 1., "Installing Service Entrance Cable and Meter Base."

b. Installing Service Entrance Wiring in Thinwall Conduit (EMT)

The Code refers to this type of conduit as Electrical Metallic Tubing (Figure 272). It is popularly known as Thinwall or EMT. Sizes of EMT used in residential wiring are the same as for rigid conduit. As the name indicates, the walls are very thin and cannot be threaded. Instead, special pressure type or setscrew connectors are used to fasten the tubing to the boxes. Setscrew connectors cannot be used outside where exposed to weather. Couplings are used to join lengths of EMT (Figure 273).

Procedures for installing EMT for the service entrance are practically the same as for rigid. **The major difference is in the type of connectors and cou-** **plings used.** Proceed as follows to install EMT using weathertight pressure connectors (Figure 274).

1. *Install the unthreaded end of the pressure connector on the end of the conduit.*
2. *Insert the connector into the meter base opening.*
3. *Screw the locknut on the connector inside the box until tight.*
4. *If using a setscrew coupling indoors to join two pieces of conduit, insert each end of conduit into the coupling to the stops. Turn the setscrews down firmly to lock the conduit to the coupling (Figure 275).*

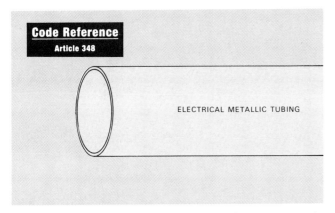

FIGURE 272. Electrical metallic tubing, commonly called thinwall or EMT.

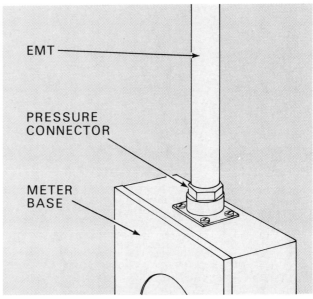

FIGURE 274. Weathertight pressure connectors are used for EMT service entrance installations.

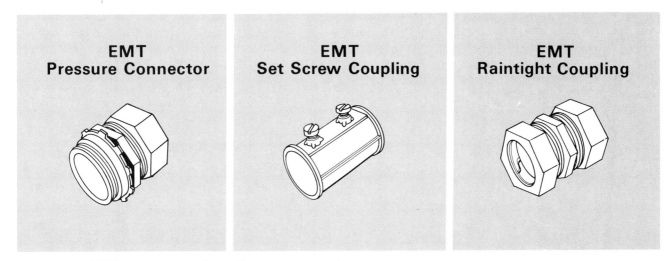

FIGURE 273. EMT connectors and couplings.

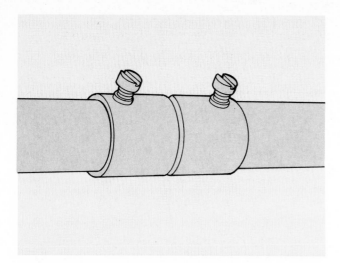

FIGURE 275. Setscrew EMT couplings may be used inside where protected from weather.

5. *Proceed as for rigid conduit to assemble conduit and install service entrance wiring.*

3. INSTALLING A MAST TYPE OF SERVICE ENTRANCE

On single-story houses with low roof lines, a rigid conduit mast is the preferred method of installing the service entrance. The rigid conduit mast extends through the roof to give the required clearance above ground (Figure 276). Service drop clearance above the roof itself may be only 46 cm (18 inches) if no more than 1.2 m (4 feet) of roof area has to be crossed by the service drop (Figure 277). Conduit size should be at least 5.1 cm (2 inches) to provide the necessary rigidity.

Masts are available in kit form which includes conduit, mounting brackets, roof flashing, entrance head and a stand-off insulator that clamps to the mast. Mast height extending more than 0.9 m (3 feet) above the roof should be supported by guy wires.

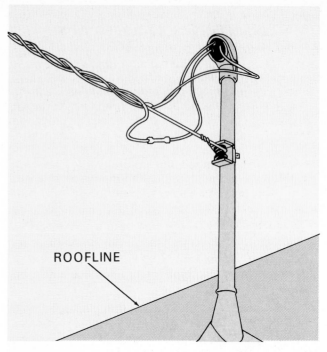

FIGURE 276. The conduit mast-type service entrance is preferred for single-story homes with low roof lines.

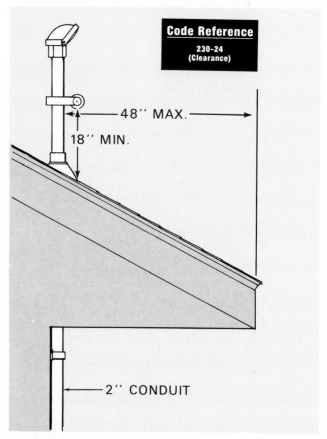

FIGURE 277. Clearance for the service entrance may be as little as 46 cm (18 inches) if no more than 1.2 m (four feet) of roof area is crossed by the service drop.

Proceed as follows to install (Figure 278):

1. *Locate meter at 1.4 to 1.5 m (4½ to 5 feet) height.*
2. *Measure to determine length of conduit needed.*

   Measure from top of meter box to required height above roof line. Assume 0.8 to 0.9 m (2½ to 3 feet), for example, above roof line to allow clearance for drip loop and weatherhead.
3. *Cut conduit to measured length.*
4. *Cut necessary holes through boards and roofing.*
5. *Install necessary bolts for bracket supports.*
6. *Insert conduit through holes and raise to position.*
7. *Fasten in place with brackets.*

8. *Install roof flashing and seal.*
9. *Install insulator for service drop (May be furnished and installed by power supplier).*
10. *Install service head as in previous procedures.*
11. *Install meter base as for conduit.*

    Install offset fitting on the conduit end to connect conduit to meter base (Figure 278). Adjust the meter base to correct position and fasten in place on the wall. Connect meter base to conduit fitting.

4. INSTALLING UNDERGROUND SERVICE ENTRANCE CABLE

Underground service is gaining favor as a means of service entrance installation. Many subdivisions now require it and more farmers are using it.

Among the advantages of underground installation are reduced danger of damage from lightning, wind, ice and falling branches. On farms it eliminates danger of tall moving equipment coming in contact with overhead wires.

In some localities, the entire distribution system is underground and the transformers are mounted on a concrete pad or fiberglass mounting unit (Figure 279). In others, only the wiring from the power sup-

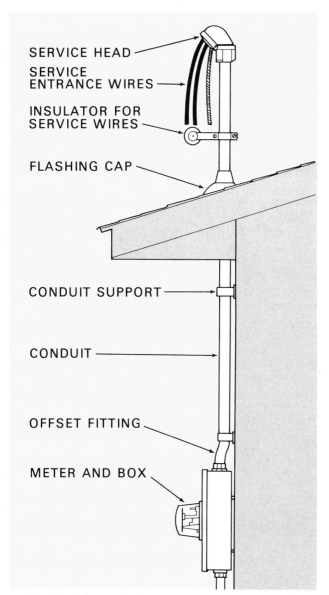

FIGURE 278. Complete mast-type service installation. An off-set fitting is installed on the conduit end to be connected to the meter base.

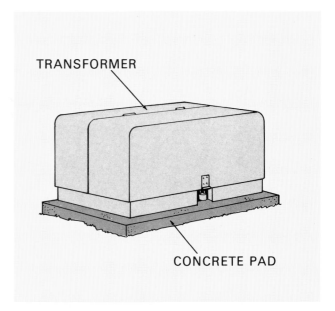

FIGURE 279. Pad-mounted transformers are used in many subdivisions for underground distribution.

135

plier overhead lines to the house is underground (Figure 280). For both types, cable is laid loosely in a trench and covered 60 to 120 cm (24–48 inches) deep, depending on local codes. If there is danger that the cable might be damaged, run it through conduit.

The cable to be installed is called Underground Service Entrance cable. The identification on the cable is Type USE (Figure 281). The conductors inside the cable are insulated with moisture-resistant insulation and the cable covering is a tough moisture-resistant material. If cable is supplied by the power supplier, as is sometimes done, it may not be USE. Most popular in some areas is single-conductor cable having rubber insulation and no outer covering.

Both recessed and surface-mounted meter installations may require a standard 90° sweep elbow with bushing, in addition to the usual conduit fittings. The bushing on the end of the elbow prevents damage to Underground Service Entrance cable (Figure 282).

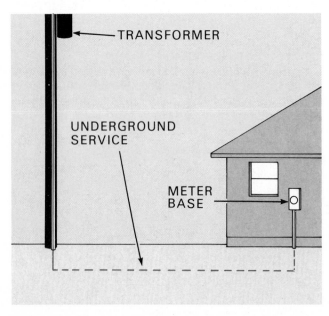

FIGURE 280. In some localities, only the wiring from the power supplier overhead lines to the house is underground.

Two methods of mounting the meter and conduit for underground service entrance are:
a. Installing Recessed Meter and Conduit.
b. Installing Surface-Mounted Meter and Conduit.

NOTE: Some power suppliers may not allow recessed meter installation. Check specifications before installing. Extra space may be required on each side and below the meter base to work on the meter.

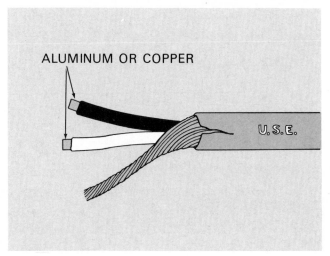

FIGURE 281. Underground service entrance cable (USE) may be buried directly in the earth.

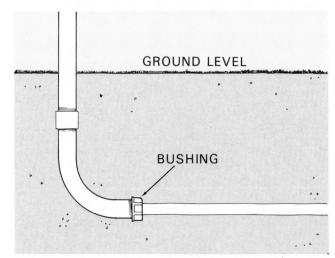

FIGURE 282. Bushings on the end of the elbow or conduit prevent damage to cable insulation.

136

a. Installing Recessed Meter and Conduit

Only rigid galvanized conduit is allowed by most codes. It is usually 5.1 cm (2-inch size). The recessed system requires installation before construction is complete, because the conduit must be mounted inside the wall space.

Proceed as follows to install conduit for underground service entrance with recessed meter and conduit (Figure 283):

1. *Dig hole for conduit installation.*

   The conduit, when installed, extends underground, so you must dig the hole to the required depth to meet the trench from the power supplier transformer. Required depth may vary from 60 to 120 cm (24 to 48 inches), depending on local codes.

2. *Mount the meter base in the wall space on the inside wall.*

   The **base** should be flush with the outside wall. The **meter** will extend from the wall when mounted.

3. *Measure from bottom of meter base to required depth underground, including the standard elbow (Figure 283).*

4. *Cut conduit to measured length, ream and thread each end.*

5. *Attach conduit to meter base and screw coupling on bottom end of conduit.*

   Use grounding type insulating bushing on end of conduit in the meter base to insure grounding of conduit (Figure 284).

6. *Screw elbow to coupling and attach bushing. Turn elbow so it is perpendicular to building when tight.*

   Power supplier personnel will install underground cable and connect to meter base terminals.

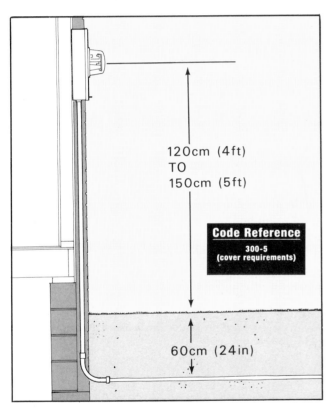

FIGURE 283. Recessed meter and underground installation.

120cm (4ft)
TO
150cm (5ft)

Code Reference
300-5
(cover requirements)

60cm (24in)

FIGURE 284. Insulating bushing with grounding lug at conduit connection in meter base.

b. Installing Surface-Mounted Meter and Conduit

Procedures for installing the surface-mounted meter base are practically the same as for recess mounted. Proceed as follows (Figure 285):

1. *Dig hole as for recess mount.*
2. *Mount meter base on outside wall surface.*
3. *Follow steps 3–6 for recess mount.*
4. *Attach mounting straps (Figure 285).*

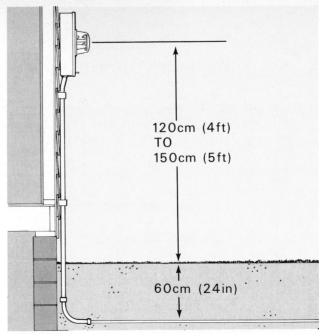

FIGURE 285. Surface-mounted meter and underground installation.

E. Connecting Service Entrance Wiring

The interior of the service entrance cabinet includes the main breaker or fuse pullout, the neutral bar and the circuit breaker panel. A grounding bar may also be installed and is required in some types of wiring installations. For a two-wire with ground system, the neutral bar must be bonded to the grounding bar (Figure 286). Instructions for installing parts of the SEP are provided by the manufacturer. Follow instructions to mount the main breaker, the neutral bar, grounding bar if present, and circuit breaker panel in the cabinet. Two to four bolts usually fasten each piece. This is usually done after all wiring has been installed in the house and circuit wires pulled into the empty cabinet.

The neutral bar and grounding bar may be mounted below the circuit breaker panel, on either side or above it, depending on the manufacturer (Figure 286). The main breaker, or fuse pullout, which disconnects all electricity to the house, is usually mounted above the circuit breaker panel (Figure 286). Or, it may be an integral part of the panel.

According to electrical equipment manufacturers, **overheating** may be a problem in SEP's after they are completely wired and placed in service. You can reduce the amount of heat generated by making SEP wiring connections in a neat, orderly manner. Excess wire length and poor connections are said to be major causes of heat generation. Cut off any excess wire before connecting to the panel and make sure all connections are tight. Without the excess length, it is easier to route circuit wires around the edges of the cabinet, allowing better air circulation (Figure 287). Shorter wires also reduce the amount of resistance. Lower resistance means less heat will be produced.

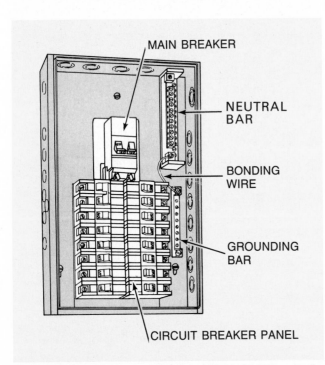

FIGURE 286. Service entrance cabinet contains main breaker, neutral bar, circuit breaker panel, and sometimes a grounding bar.

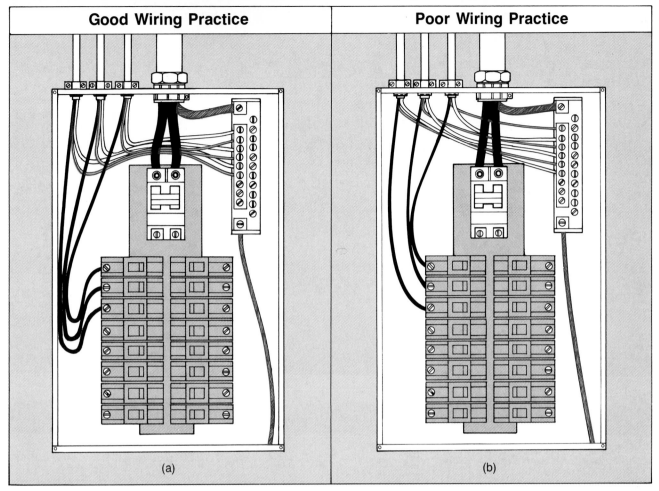

| Good Wiring Practice | Poor Wiring Practice |
| (a) | (b) |

FIGURE 287. Remove excess wire length for easier installation and better air circulation.

Compare the two panels in Figure 287. Wiring on the left panel is the ideal. On the right is a poor wiring job. Each panel has the same number and ratings of breakers.

Connecting SEP wiring is discussed as follows:
1. Connecting Service Entrance Conductors to Service Entrance Panel Terminals.
2. Installing Wiring System Ground.

1. CONNECTING SERVICE ENTRANCE CONDUCTORS TO SERVICE ENTRANCE PANEL TERMINALS

Proceed as follows:

1. *Connect the two hot conductors (Figure 287).*

   The hot conductors usually are either two black wires or one red and one black. Blue and other colors may be used for hot wires. Cut off any excess length on the service entrance wires. Insert each wire into the connector provided in the main breaker terminal. Tighten the terminal screws. Many are hexagonal and require use of an "Allen" wrench of the correct size. Others are slotted for regular screwdrivers. **If using aluminum wire, apply inhibitor paste to each terminal and conductor. Be sure the terminals are labeled Al-Cu. Make every connection as tight as practicable.**

2. *Connect the neutral wire to the neutral bar terminal.*

   Apply inhibitor and tighten securely with an Allen wrench, or screwdriver.

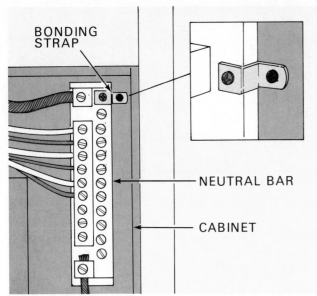

FIGURE 288. Bonding strap must be bolted to neutral bar and to cabinet.

3. *Ground and bond neutral bar to the metal panel cabinet.*

Use bolt or strap provided (Figure 288). Insert bonding screw through the hole provided in the neutral bar and tighten. If a bonding strap is provided, bolt one end to the neutral bar and the other to the metal cabinet. If the cabinet is non-metallic, the terminal bar must be connected to the grounding conductor that is run with the service entrance conductors. Grounding conductors must not be connected to the terminal bar provided for the grounded conductors unless the bar is identified for the purpose. Bonding the cabinet to the neutral bar provides additional grounding protection against ground faults that may occur on the inside or outside surface of the metal cabinet.

NOTE: The screw used to bond the neutral bar to a metal cabinet must be green and must be visible after installation, according to Code Section 250-79.

2. INSTALLING WIRING SYSTEM GROUND

To ground the house wiring system, you must connect a **ground wire** (grounding electrode conductor) from the neutral bar on the panel board to an approved grounding electrode, such as a metal water pipe and/or to other approved grounding electrode(s). Where a metallic water pipe is used as a grounding electrode the Grounding Electrode

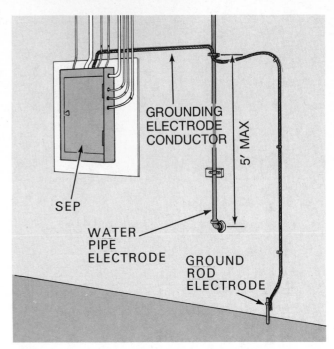

FIGURE 289. If metal water pipe is used as a grounding electrode, it must be supplemented by one or more additional electrodes.

Conductor (GEC) must be connected to the electrode within the first 5 feet of entering the structure. **If metal water pipe is used, it must be supplemented by one or more additional electrodes to meet Code requirements** (Figure 289).

a. Sizes and Types of Grounding Conductors

Sizes and types of grounding conductors are discussed under the following headings:

(1) Sizes of Grounding Electrode Conductors.
(2) Types of Grounding Electrode Conductors.

(1) SIZES OF GROUNDING CONDUCTORS. The size of the grounding conductor depends on the size of your service entrance conductors. **You may use No. 8 copper wire if your service entrance conductors are No. 2 or smaller.** However, you may want to use No. 6 instead, because it is simpler to install. For example, a No. 8 ground wire must be enclosed in conduit or armored cable. A No. 6 or larger wire may be stapled rigidly to the building surface without being covered unless exposed to severe physical damage (Figure 290). Select grounding conductor size from Table XII (taken from Code Table 250–94).

The grounding electrode conductor must be of copper, aluminum or copper-clad aluminum.

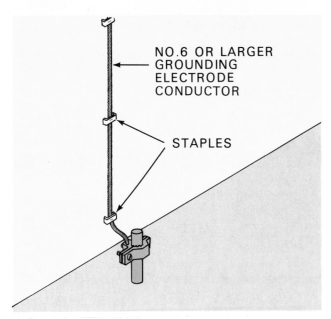

FIGURE 290. Number six or larger grounding electrodes may be stapled rigidly to the wall without being covered unless it is exposed to severe physical damage.

(2) TYPES OF GROUNDING CONDUCTORS.
Aluminum or copper-clad aluminum may not be used in contact with masonry or the earth or in corrosive conditions. If used outside, it may not be installed within 46 cm (18 inches) of the earth (see Code Section 250-92 [a]).

TABLE XII. SIZES OF GROUNDING ELECTRODE CONDUCTORS FOR AC SYSTEMS
(From Code Table 250-94)

Reprinted with permission from NFPA 70–1993, the National Electrical Code® , Copyright 1992, National Fire Protection Association, Quincy, MA 02269. This reprinted material is not the complete and official position of the National Fire Protection Association, on the referenced subject which is represented only by the standard in its entirety.

Size of Largest Service-Entrance Conductor or Equivalent Area for Parallel Conductors		Size of Grounding Electrode Conductor	
Copper	Aluminum or Copper Clad Aluminum	Copper	Aluminum or Copper Clad Aluminum*
2 or smaller	1/0 or smaller	8	6
1 or 1/0	2/0 or 3/0	6	4
2/0 or 3/0	4/0 or 250 kcmil	4	2
Over 3/0 thru 350 kcmil	Over 250 kcmil thru 500 kcmil	2	1/0
Over 350 kcmil thru 600 kcmil	Over 500 kcmil thru 900 kcmil	1/0	3/0
Over 600 kcmil thru 1100 kcmil	Over 900 kcmil thru 1750 kcmil	2/0	4/0
Over 1100 kcmil	Over 1750 kcmil	3/0	250 kcmil

Where multiple sets of service-entrance conductors are used as permitted in Section 230–40, Exception No. 2, the equivalent size of the largest service-entrance conductor shall be determined by the largest sum of the areas of the corresponding conductors of each set.

Where there are no service-entrance conductors, the grounding electrode conductor size shall be determined by the equivalent size of the largest service-entrance conductor required for the load to be served.

*See installation restrictions in Section 250–92(a).

(FPN): See Section 250–23(b) for size of alternating-current system grounded conductor brought to service equipment.

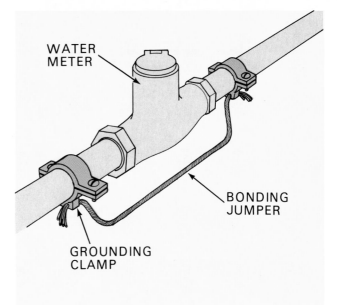

FIGURE 291. Run a bonding jumper around a water meter or insulated sections of pipe to assure grounding continuity. A jumper should be long enough to permit removal of the meter without breaking the bond.

b. Grounding Electrodes

Bonding is important in installing grounding electrodes. It is especially important in meeting Code requirements.

Bonding means the permanent joining of metallic parts to form a conductive path for electrical current that will assure electrical continuity and capacity to conduct safely any current likely to be imposed on it. The bonding jumper around the water meter in Figure 291 is an example.

Whether used as part of the grounding electrode system or not, the interior metal water piping system must always be bonded to either the service entrance cabinet, the grounding electrode conductor or the grounding electrode. Also, other metal piping which may become energized, such as gas pipes, must be bonded. A metal underground gas piping system must not be used as a grounding electrode. The size of the bonding conductor must be the same as the grounding electrode conductor.

Grounding electrodes are discussed under the following headings:
(1) Bonded Electrode Systems.
(2) Made and Other Electrodes.

(1) BONDED ELECTRODE SYSTEMS (Code Section 250-81). Bonded electrode systems are discussed as follows:
(a) Four-Electrode Bonded Systems.
(b) Three-Electrode Bonded Systems.

(a) Four-electrode bonded systems are discussed as follows:

141

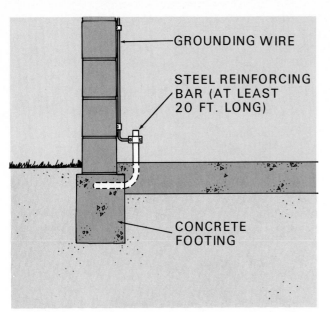

FIGURE 292. Half-inch (1.3 cm) steel reinforcing bars or No. 4 copper wire at least 6.1 m (20 feet) long may be used as supplemental grounding electrodes if encased by 5.1 cm (2 inches) of concrete in direct contact with the earth.

Under the Code, each of the following four items, if available, must be bonded together to form the grounding electrode system.
—A **metal underground water pipe** in direct contact with the earth for at least 3.1 m (10 feet).
— The **metal frame of the building** (not applicable to most homes).
—An **electrode encased in at least 5.1 cm (2 inches) of concrete** at the bottom of the footing consisting of at least 6.1 m (20 feet) of one or more 1.3 cm (½-inch) steel bars or rods, or bare copper at least No. 4 AWG.
—A **ground ring** of bare copper not smaller than No. 2 AWG encircling the building, buried at least 76 cm (2½ feet) deep.

Note that the code does *not* state that all of the systems described **must be present,** only that they must be bonded together if available.

If you connect to the metal water pipe for the grounding electrode, the pipe must be in direct contact with the earth for at least 3.1 m (10 feet) (a metal well casing may be used). If there is a water meter or insulated joints or sections in the line that may be removed for repairs or replacement, you must run a jumper around it, using the same size conductor as the ground wire (Figure 291). Where used as a grounding electrode, a metal underground water pipe must be supplemented by an additional electrode as previously described.

(b) Three-electrode bonded systems are discussed as follows:

The metal underground water pipe must be bonded to both of the following electrodes **where available**:

—An **electrode** encased by at least 5.1 cm (2 inches) of concrete in direct contact with the earth and at least 6.1 m (20 feet) long (Figure 292). If steel reinforcing bars are used, they must be at least 1.3 cm (½ inch) in diameter. Bare solid or stranded copper wire No. 4 or larger may be used.
—A **ground ring** encircling the building in direct contact with the earth buried at least 76 cm (2½ feet) deep, made of bare copper wire No. 2 or larger, at least 61. m (20 feet) long (Figure 292A). Where none of the above electrodes are available, one or more of the following types may be used.

A typical connection of a grounding electrode system in a residence is shown in Figure 293A.

(2) MADE AND OTHER ELECTRODES. Made electrodes are those installed using metal rod, pipe or plate.

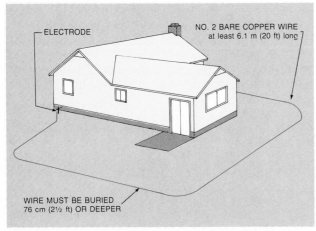

FIGURE 292A. Ground ring of bare copper wire, No. 2 or larger, at least 6.1 m (20 feet) long must be buried at least 76 cm (2½ feet) deep.

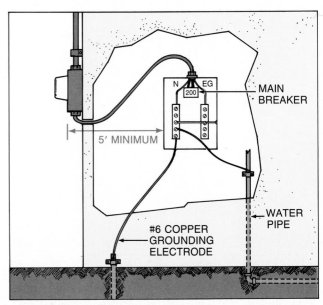

FIGURE 293A. The size of the GEC is based on the size of the service entrance conductors.

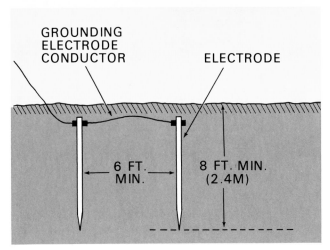

FIGURE 293B. Top end of electrode must be flush with ground level or below, if unprotected. Two or more rods must be at least six feet apart.

Made electrodes must be free of paint or enamel and buried below permanent moisture level where practicable. They are discussed as follows:

— **Rod and pipe made electrodes** must not be less than 2.4 m (8 feet) in length and pipe shall be 1.9 cm (¾ inch) trade size or larger. The grounding electrode conductor for made electrodes must be No. 6 copper or No. 4 aluminum. They must be galvanized or metal coated for corrosion protection (some local codes do not permit the use of galvanized pipe).

The electrode must be installed so that at least 2.44 m (8 feet) of length is in contact with the soil.

The electrode, rod or pipe, shall be driven to a depth of 2.4 m (8 feet) unless rock bottom in encountered. If you hit rock bottom at less than 1.2 m (4 feet), you must bury the electrode in a trench 762 mm (2½ feet) deep, or drive it at an oblique angle not to exceed 45 degrees from vertical. The top end of the electrode must be flush with or below ground level unless it and the grounding electrode conductor are both protected against physical damage. Where multiple rod, pipe or plate electrodes are installed, they must be at least 1.8 m (6 feet) apart (Figure 293B).

— **Steel or iron rods** must be at least 1.6 cm (⅝-inch) diameter. Non-ferrous rods shall be listed and not less than 1.3 cm (½-inch) diameter.

Stainless steel rods may be 1.3 cm (½-inch) in diameter.

— **Plate electrodes** of iron or steel must expose at least 0.19 square meters (2 square feet) to the soil and must be at least 6 mm (¼-inch) thick . Non-ferrous plates shall be at least 1.5 mm (0.06-inch) thick.

— **Underground metal gas piping systems shall not** be used as an electrode.

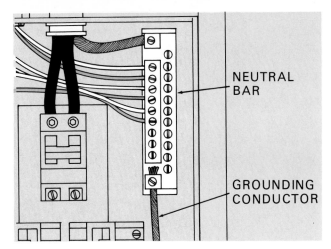

FIGURE 294. Grounding conductor connected to neutral bar.

— **Aluminum electrodes** are **not** allowed by the Code.
— Other **metal piping systems** and **underground tanks** may be approved by some local codes.

c. Connecting Grounding Conductors and Grounding Electrodes for System Ground

The connection of the grounding conductor to the grounding electrode must be accessible and made in a manner that will provide a permanent and effective ground.

Proceed as follows to connect the system ground:
1. *Connect the grounding conductor at the neutral bar.*

   Insert the grounding conductor under the terminal provided on the neutral bar and tighten securely (Figure 294).

2. *Connect the grounding conductor at the water pipe or other grounding electrode.*

   The ground wire must be clamped tightly to the water pipe or other grounding electrode. Use bronze, brass, plain or malleable iron clamps (Figure 295).

   Two steps are involved to make the connection. **First**, connect the clamp to the pipe using the two bolts provided. Tighten the bolts until the ridges in the clamp bite into the water pipe. **Second**, insert the ground wire in the slot on top of the clamp and draw the screw(s) or bolt(s) tight against the wire. The metal water piping bonding jumper must be accessible.

3. *Bond the first grounding electrode to at least one other grounding electrode.*

   Follow procedures as in step two.

143

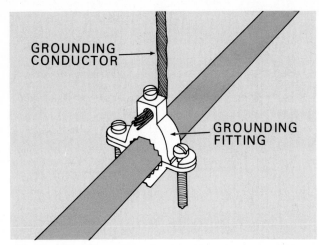

FIGURE 295. Grounding fitting clamped tightly to water pipe.

NOTE: Some power suppliers and public service authorities require the ground wire to be installed differently. It does not go through the SEP. Instead, it connects to the neutral overhead wire at the weatherhead or at the neutral bar in the meter base. It comes down the side of the house and fastens directly to the water pipe and/or other type of ground (Figure 296).

The Code requires that separate systems serving a home be bonded together. For example, grounding systems for cable television, telephone and burglar alarms must be interconnected to reduce differences of potential between them which can result from a power surge or lightning. If each system has a separate grounding electrode, all must be bonded together as shown in figure 293B, and bonded to the electrical power system ground. The Code also requires that there must be provided on the outside of the house an

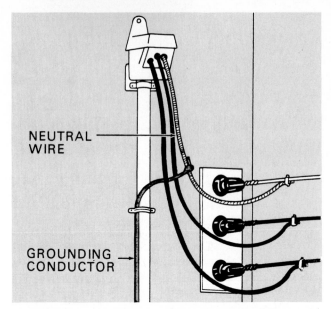

FIGURE 296. Grounding conductor connected to the overhead neutral wire and directly to a grounding electrode is required by some Codes.

accessible means of bonding such systems to the power system ground. It may be one of the following: (1) an exposed grounded conduit, (2) an exposed grounding electrode conductor, or (3) an approved means for the external connection of a bonding or grounding conductor to the service equipment.

A number 6 copper conductor connected as in figure 290 with at least 15.2 cm (6 inches) exposed on the outside wall of a house is an example of an approved means of bonding separate systems to the power system ground. A bonding jumper not smaller than number 6 copper must be connected between the separate systems and the power system ground.

F. Installing Ground Fault Circuit Interrupters (GFCI)

Ground Fault Circuit Interrupters (GFCI's) are people-protectors that have proven their worth in preventing death and serious injury from electric shock. System and equipment grounding provide for protection for the electrical system and equipment in most situations, but there is a need for a more sensitive device to protect people from shock exposure from hand-held equipment and in more hazardous locations in and around the house.

GFCI's are installed in all NEC-mandated areas to protect 15- and 20-ampere 120-volt receptacles. A leakage of 5 milliamperes, or 0.005 A of the current required to open a 15-ampere circuit breaker or fuse, can be fatal. A small 5 milliamperes of leaking current will trip open a GFCI. This amount of current is not enough to cause death.

GFCI's protect personnel by continually monitoring the current entering the hot conductor and comparing that current level with the amount of current returning via the grounding circuit conductor (neutral) (Figure 297).

GFCI's and their application should not be confused with **Ground Fault Interrupters (GFI)**. With a GFCI, the amount of current fault is limited to 5 milliamperes (plus or minus one mill). Where a GFI is involved, there are usually large amounts of current anticipated in a condition where phase conductors may short to each other or one phase may short to the ground. This condition, if not interrupted, will cause massive destruction to both equipment and structures.

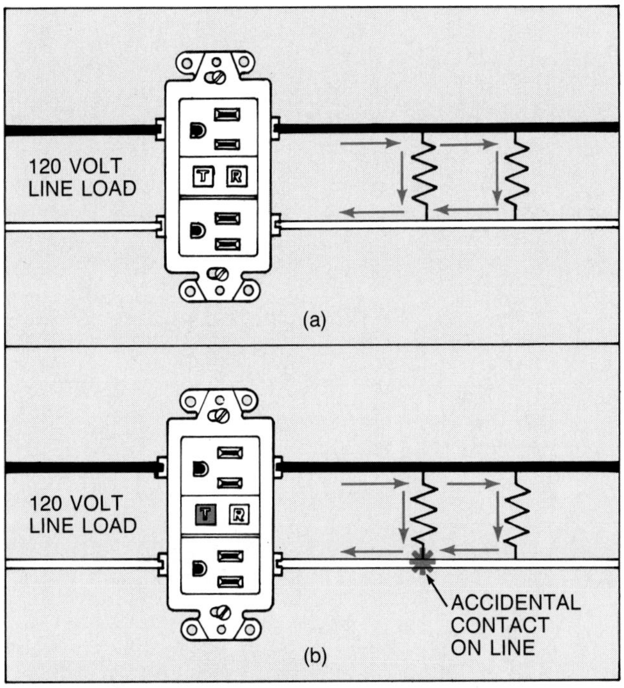

Installing GFCI's is discussed as follows:
1. Types and Sizes of GFCI's.
2. Identifying Specific Areas Where GFCI's Are Required.
3. Identifying Additional Areas Where GFCI's are Required.
4. Connecting GFCI's.
5. Testing GFCI's.

1. TYPES AND SIZES OF GFCI's

Three types of GFCI's are available. **Portable GFCI's** are intended as an extension to outlets and are intended for temporary use such as needed at a construction site (Figure 298).

FIGURE 298. Portable GFCI's can be located where needed on construction sites.

```
120 VOLT
LINE LOAD

(a)

120 VOLT
LINE LOAD

ACCIDENTAL
CONTACT
ON LINE
(b)
```

FIGURE 297 (a). A GFCI-monitored receptacle with equal current in the black wire and the white lead will continue to operate. (b) A 5 mil imbalance in current will cause the GFCI to trip, removing voltage to load.

145

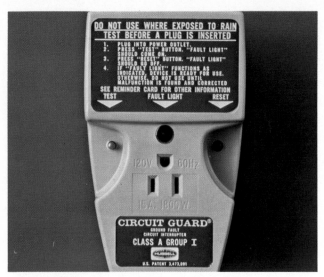

FIGURE 299. A portable GFCI that plugs directly into an existing receptacle.

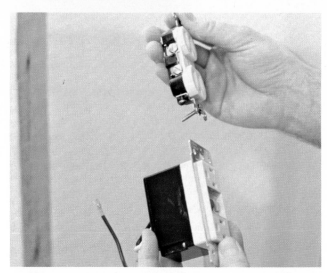

FIGURE 301. A GFCI (bottom) can be permanently installed in place of a receptacle.

Many of these portable units plug into an existing outlet (Figure 299). These outlets are generally protected in the conventional manner with a 15- or 20-ampere breaker. Plugging a portable GFCI into an extension cord will provide people protection to a degree equal to plugging equipment into a permanently installed GFCI.

A **breaker-style GFCI** can replace an existing breaker and snap directly into the panelboard stabs (Figure 300). Generally, they are 20-ampere rated and provide the same equipment protection as do conventional breakers. Breaker-type GFCI's contain overcurrent overload and GFCI sensing (Figure 300).

Receptacle type GFCI's are the most popular type of protectors. They are available in a variety of colors and sizes and cost considerably less than the breaker type (Figure 301).

A receptacle-type GFCI has two distinct and separate sets of connection terminals, one marked "line terminals" and the other marked "load terminals." The second set of terminals marked "load" allows greater utilization of this input GFCI to protect additional receptacles connected downstream (Figure 302). There are restraints when this type of connection is considered since the coupling that exists between the conductors allows a certain amount of current leakage. This is a natural happening and does not mean there is anything defective concerning your branch circuit. Care should be taken not to connect too many downstream outlets which would increase conductor lengths and cause nuisance tripping of the GFCI.

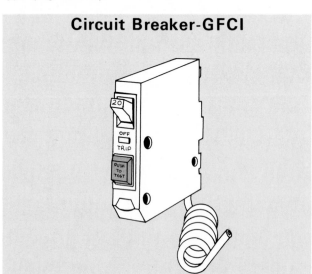

Circuit Breaker-GFCI

FIGURE 300. The combination circuit-breaker GFCI is installed in the service entrance panel. It protects all outlets on that circuit.

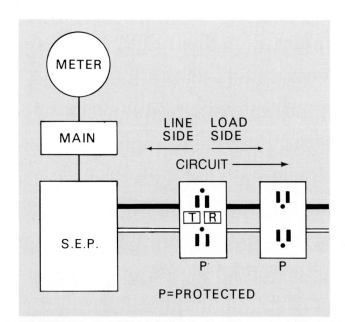

FIGURE 302. A GFCI installed as the first receptacle in a circuit provides protection to all other receptacles downstream (beyond it on the current).

146

Of course, the question is how many outlets may be connected to a GFCI circuit. Unfortunately, there aren't any easy answers since conductor lengths vary considerably between outlets. When planning for outlets, equal consideration should be given to conductor lengths with the number and placement of outlets. It is possible that nuisance tripping can occur with a few outlets if the distance between outlets is considerable.

2. IDENTIFYING SPECIFIC AREAS WHERE GFCI'S ARE REQUIRED

GFCI installation requirements have been expanded for the last several Code cycles. The 1993 edition of the Code lists several new requirements. A complete list of NEC-required residential GFCI installations and exceptions are listed in Section 210-8 of the 1993 Code.

Identifying areas where GFCI's are required is discussed under the following headings:

a. Bathroom Requirements of GFCI's.
b. Garage Requirements for GFCI's.
c. Outdoor Area Requirements for GFCI's.
d. Crawl Space and Unfinished Basement Requirements for GFCI's.
e. Kitchen Counter Requirements for GFCI's.
f. Boathouse Requirements for GFCI's.

a. Bathroom Requirments for GFCI's (Code Section 210-8a [1])

According to the Code, all receptacles in a bathroom must be protected by a GFCI. A bathroom is defined as an area including a basin and one or more of the following: a toilet, a tub or a shower.

b. Garage Requirements for GFCI's (Code Section 210-8a [2])

All 120-volt, 15- and 20-ampere receptacles in garages must be GFCI-protected, with two exceptions (Figure 303).

A GFCI is not required for an overhead door receptacle, since it is considered not to be readily accessible.

The other exception is for outlets considered not readily accessible because of their location behind a freezer or similar equipment, provided all openings are occupied by a plug attachment.

If there is not a two-plug attachment made to a duplex receptacle, or if one side of the receptacle is not occupied by a plug attachment, then that receptacle is

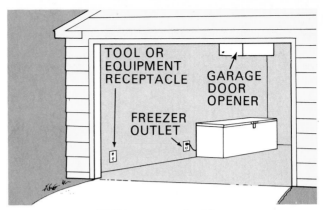

FIGURE 303. A GFCI is not required for a garage receptacle in a space provided for a plug-in appliance or a garage door opener.

not exempt from the GFCI requirements. In this case, a single receptacle would be required. Since there would be no unused outlets, the receptacle would then be exempt from GFCI requirements.

c. Outdoor Area Requirements for GFCI's (Code Section 210-8a [3])

All outdoor receptacles installed within 6'6" of final grade are considered to be accessible and require GFCI protection.

d. Crawl Space and Unfinished Basement Requirements for GFCI's (Code Section 210-8a [4])

With three exceptions, all receptacles installed in a crawl space or below grade and those in unfinished basements, must be GFCI-protected. The exceptions are as follows:

1. A single receptacle dedicated and identified for refrigerators, freezers, or other fixed appliances. A duplex circuit may **not** be installed **unless** both sides are occupied with a plug attachment and still meet the guidelines of this exception.
2. Laundry circuits.
3. A single receptacle supplying sump. There are no provisions allowing a duplex receptacle to be installed for this use under this exemption.

147

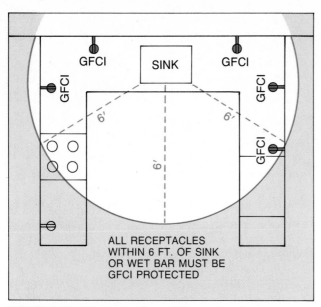

FIGURE 304. All receptacles within 6' of a sink or wet bar must be GFCI-protected.

e. Kitchen Counter Top Surface Requirements for GFCI's
(Code Section 210-8a [5])

All 15- and 20-ampere, 120-volt receptacles serving counter top surfaces within 1.83m (6 feet) of a kitchen sink or wet bar must have GFCI protection (Figure 304). Kitchen receptacles allowed by Code Section 422-8 (d) are not required to have GFCI protection.

f. Boathouse Requirements for GFCI's
(Code Section 210-8a [6])

The Code requires that all 15- and 20-ampere, 120-volt receptacles installed in a boathouse be provided with GFCI protection.

3. IDENTIFYING ADDITIONAL AREAS WHERE GFCI'S ARE REQUIRED

There are areas identified throughout the NEC where GFCI protection is mandated. A few of these areas of particular importance to the residential environment are: permanently installed pools, fountains, storable swimming or wading pools, and construction sites.

When receptacles are replaced, the NEC requires that the replacement device meet or exceed the present Code. For the first time an installer must bring all replaced receptacles up to current code requirements. It is no longer possible to exchange like type for like type where the current Code requires a GFCI. See Code Section 210-7 (d) for more detailed information.

Additional provisions were made where non-grounding type receptacles are to be replaced with a grounding type receptacle without the grounding pin being connected to the equipment grounding system, provided the receptacle is protected by a GFCI. This is an accepted practice where there is no equipment grounding conductor present.

All new construction will require the equipment grounding conductor to be installed and connected to the grounding pin of a 3-prong receptacle. Where the grounding pin does not have an equipment grounding conductor connected, the receptacle must be indentified as being protected by a GFCI (Figure 305). All GFCI packages contain stick-on identifying labels for this purpose.

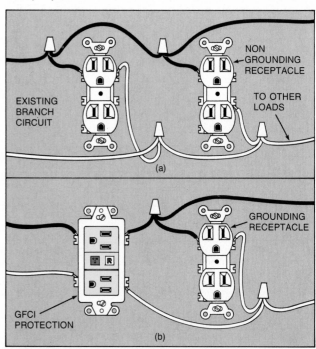

FIGURE 305. Where the grounding pin does not have an equipment grounding conductor connected, the receptacle must be identified by a stick-on label as being protected by a GFCI.

4. CONNECTING GFCI'S

Connecting GFCI's is discussed under the following headings:
  a. Connecting the Plug-In Type GFCI.
  b. Connecting Permanently Installed Receptacle Type GFCI.
  c. Connecting the Combination Circuit Breaker Type GFCI.

a. Connecting the Plug-In Type GFCI

1. *To connect the plug-in type, plug the device directly into the receptacle.*

b. Connecting the Permanently Installed Receptacle Type GFCI

To connect the permanently installed receptacle-type GFCI in a circuit, follow manufacturer's directions. Proceed as follows:

1. *Install the GFCI in a standard switch (receptacle) box.*
2. *Connect wiring as for standard receptacle, white to white terminal, black to brass terminal and ground to ground (some types are self-grounding).*
3. *If desired, make feed-through connections to other receptacles.*

   Most types have terminals for continuing to other outlets, which then are also protected. They provide GFCI protection at the point of installation and at all other receptacles connected beyond the unit on the same circuit.
4. *Mount the outdoor type as directed by manufacturer's instructions.*

c. Connecting the Combination Circuit Breaker Type GFCI

To connect circuit breaker type GFCI in service entrance panel, follow manufacturer's instructions. Most will be connected as illustrated.

Proceed as follows to connect the combination circuit breaker and GFCI (Figure 306):

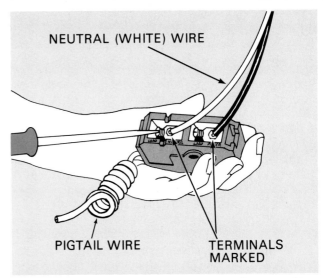

FIGURE 306. Connect white circuit wire to the neutral terminal and black circuit wire to other terminal on circuit breaker GFCI.

1. *Connect the black wire to terminal marked load power.*
2. *Connect white neutral wire from circuit to screw terminal on GFCI marked load neutral.*
3. *Plug the device into the SEP clips.*
4. *Connect the white pigtail wire to the neutral bar (Figure 306).*

5. TESTING GFCI'S

Each GFCI has a test button that you can use to determine if the device is working (Figure 306A). Owners should test each unit monthly by pushing the test button. If the unit is working, it will trip the circuit breaker on the circuit. Reset the circuit breaker after the test.

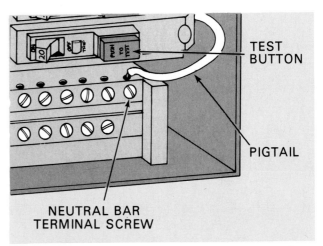

FIGURE 306A. Test the GFCI monthly according to directions on the units.

147B

G. Connecting Circuits in Service Entrance Panel

As illustrated in Figure 287, an orderly wiring job in the SEP allows better air circulation, reducing heat buildup. Another advantage is that it is much easier to identify circuit wiring and breakers when troubleshooting and for adding circuits at a later date if desired.

You will install a single-pole circuit breaker for each 120-volt circuit and a double-pole one for the 240- or 120/240-volt circuits (Figure 306). The capacity of the circuit breaker is rated in amperes and is marked on the lever. You must match the rating of the circuit breaker to the amperage of the circuit it protects.

The ends of each circuit cable should have already been pulled into the service entrance cabinet and identified. Check each cable before you make the connection. The code requires that you mark the identification of the circuit on the chart inside the cabinet door. The markings must be legible and durable.

For each cable connection, strip the sheath from the cable. Extend the wires to the terminals to which they connect and cut off any excess length. Then remove about 1.3 cm (½-inch) of insulation from each individual wire.

Connecting circuits in the SEP is discussed as follows:

1. Connecting 120-Volt Circuits with Circuit Breakers.
2. Connecting 240-Volt Circuits with Circuit Breakers.
3. Connecting 120/240-Volt Circuits with Circuit Breakers.
4. Connecting 120-Volt Circuits with Fuses.
5. Connecting 240-Volt Circuits with Fuses.
6. Connecting 120/240-Volt Circuits with Fuses.

1. CONNECTING 120-VOLT CIRCUITS WITH CIRCUIT BREAKERS

Proceed as follows:

1. *Extend the white wire and the bare grounding wire around the edge of the cabinet to the neutral bar (Figure 307).*

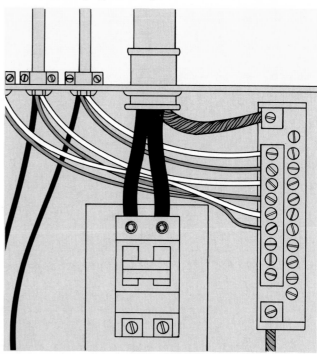

FIGURE 307. White neutral wire and grounding wire attached to terminals on neutral bar.

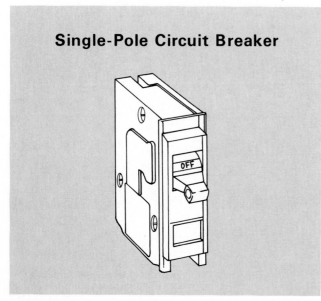

Single-Pole Circuit Breaker

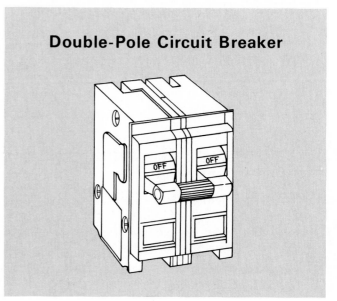

Double-Pole Circuit Breaker

FIGURE 306. Single-pole and double-pole circuit breakers.

2. *Insert the white wire under a neutral bar terminal and tighten the screw firmly.*

3. *Insert the bare grounding wire under a neutral bar terminal and tighten the screw.*

4. *Insert the black wire from the cable under the terminal screw on the circuit breaker and tighten the screw (Figure 308).*

5. *Insert the circuit breaker into the slot at one end and push the blade into the clip terminal until it stops (Figure 309).*

Procedures for some makes may vary slightly, but most will plug into the terminal.

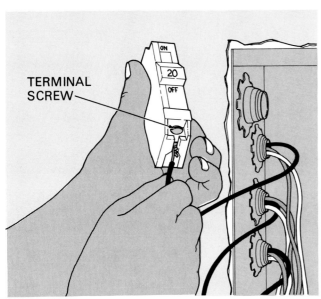

FIGURE 308. Black ("hot") wire inserted into circuit breaker terminal.

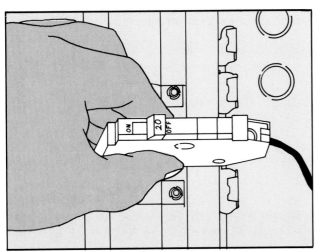

FIGURE 309. Circuit-breaker snaps into position on the panel.

To remove a circuit breaker after installation, insert a screwdriver blade under the end opposite the wire connection and pry up until the clip releases.

CAUTION: When removing a circuit breaker from a SEP where power is turned on, be sure to turn off power at main disconnect before removing breaker.

2. CONNECTING 240-VOLT CIRCUITS WITH CIRCUIT BREAKERS

Certain equipment such as water heaters, 240-volt air conditioners and 240-volt heaters are installed on straight 240-volt circuits, not 120/240-volt as with a range or dryer. These 240-volt circuits are wired with 2-wire w/g cable as follows (Figure 310):

1. *Connect black wire to a terminal on the 240-volt circuit breaker.*

2. *Connect white wire to the other breaker terminal (paint the end of white wire black, or wrap with tape to identify as hot wire).*

3. *Connect the grounding conductor to the neutral bar.*

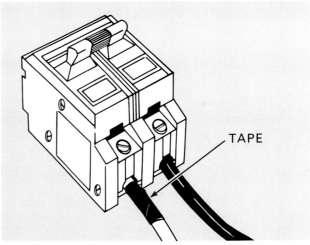

FIGURE 310. Double-pole circuit breaker with 240 circuit wires attached.

3. CONNECTING 120/240-VOLT CIRCUITS WITH CIRCUIT BREAKERS

1. *Extend the white neutral wire from the cable around the edge of the cabinet to the neutral bar.*
2. *Insert the white neutral wire under a neutral bar terminal and tighten the screw.*
3. *Insert one hot wire (black) under a terminal on a double-pole circuit breaker and tighten (Figure 311).*
4. *Insert the other hot wire (red) under the other terminal on the double-pole circuit breaker and tighten.*
5. *Insert grounding wire under a neutral bar terminal and tighten screw.*

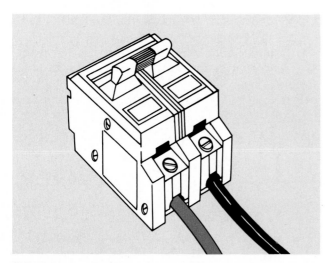

FIGURE 311. Double-pole circuit breaker with 120/240 circuit wires attached.

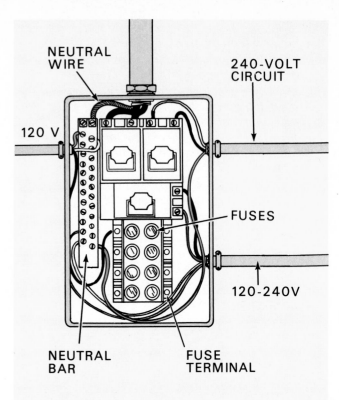

FIGURE 312. Fuse panel with 240-volt circuit, 120/240 volt circuit and 120-volt circuit installed.

4. CONNECTING 120-VOLT CIRCUITS WITH FUSES

Procedures are similar to those for circuit breaker wiring. Proceed as follows (Figure 312):

1. *Extend the white neutral wire and the bare grounding wire to the neutral bar.*
2. *Insert the white neutral wire under a neutral bar terminal and tighten.*
3. *Repeat for the bare grounding wire.*
4. *Extend the black wire to a fuse terminal.*
5. *Insert the black wire under the terminal screw and tighten.*

5. CONNECTING 240-VOLT CIRCUITS WITH FUSES

Proceed as follows (Figure 312):

1. *Connect the black wire to a 240-volt fuse terminal.*
2. *Connect the white wire to the other 240-volt terminal and identify as a hot wire with tape or paint.*
3. *Connect grounding wire to neutral bar.*

6. CONNECTING 120/240-VOLT CIRCUITS WITH FUSES

Proceed as follows (Figure 312):

1. *Extend the white neutral wire to the neutral bar.*
2. *Insert the white neutral wire under a neutral bar terminal screw and tighten.*
3. *Extend one hot wire (black) to a fuse connection (Figure 312).*
4. *Insert the hot wire under the fuse-terminal screw and tighten.*
5. *Repeat steps 3 and 4 for the second hot wire (red).*
6. *Insert grounding wire under a neutral bar terminal and tighten screw.*

V. Installing Metallic Conduit

Three types of metallic conduit were described in Section IV, "Installing Service Entrance Equipment." The three types are **rigid conduit**, **intermediate metallic conduit**, and **EMT (thinwall) conduit** (Figure 313). All three come in 3.1 m (10 feet) lengths and must be UL approved. Other types are **flexible metallic conduit**, often called **Greenfield liquidtight** and **flexible metal** conduit (Figure 313). The use of flexible metallic conduit is limited to dry locations, and it must be protected from physical damage if not covered. In certain areas it is used extensively, but in the United States as a whole, it is not widely used in house wiring. It is not discussed in this section.

Rigid conduit and IMC are seldom used in house wiring, except as described for service installations, and are not discussed in this section.

EMT is lighter and easier to handle and is approved for most requirements in house wiring installations. Many local codes require the use of EMT for all residential wiring. EMT is available in sizes from 1.3 to 10.2 cm (½-inch to 4 inches).

Rigid nonmetallic conduit (PVC) is economical and may be used in walls, floors and ceilings (Figure 313). It may also be used for exposed areas if the building has no more than three floors and the material is not exposed to physical damage. Liquidtight flexible metal conduit is similar to flexible metal conduit but has a jacket that is liquidtight, nonmetallic and sunlight resistant (Figure 313). It may be used for direct burial and for the exterior of buildings where flexibility is desired for protection from liquids, vapors, or solids. Its use for the installation of service entrance conductors is also allowed.

Upon completion of this section, **you will be able to install metallic conduit and the necessary wiring**.

Installing metallic conduit and wiring is discussed as follows:

A. Installing Electrical Metallic Tubing (EMT).
B. Installing Wiring in Metallic Conduit.

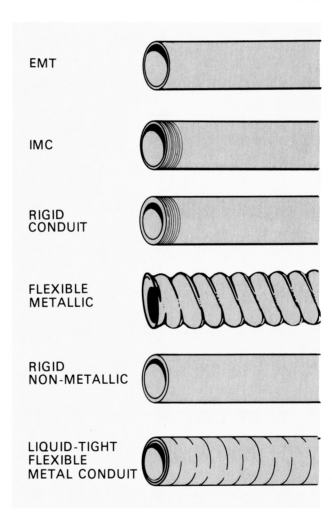

EMT

IMC

RIGID CONDUIT

FLEXIBLE METALLIC

RIGID NON-METALLIC

LIQUID-TIGHT FLEXIBLE METAL CONDUIT

FIGURE 313. Types of conduit installed for house wiring.

A. Installing Electrical Metallic Tubing (EMT)

Installing EMT is discussed under the following headings:
1. Cutting Electrical Metallic Tubing.
2. Bending Electrical Metallic Tubing.
3. Installing Electrical Metallic Tubing.

1. CUTTING ELECTRICAL METALLIC TUBING

Cut the measured length of EMT with either a fine blade hacksaw (13 teeth to the centimeter or 32 teeth to the inch) (Figure 314) or with a tubing cutter. Use a reamer to remove sharp edges after cutting.

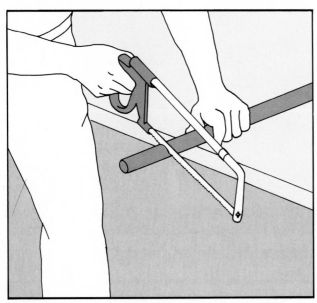

FIGURE 314. Cutting EMT with a fine-tooth hacksaw.

2. BENDING ELECTRICAL METALLIC TUBING

You must be very careful when bending EMT to avoid kinking or crushing the pipe, which would reduce the usable inside area. A run of EMT between outlets, or between fittings or between outlet and fitting must not contain more than the equivalent of four quarter bends total (360 degrees). Bend the conduit first, then cut the length required.

Thinwall conduit benders are available which are marked to help you make the type and length of bend required (Figure 315). The foot step helps to apply steady pressure during bending (Figure 315). Always work on a hard surface when bending EMT. Follow manufacturer's directions.

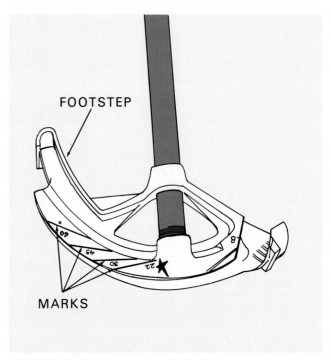

FOOTSTEP

MARKS

FIGURE 315. Marks on bender serve as a guide.

Assume you want a 90° bend, 28 cm (11 inches) high. Proceed as follows to bend 1.3 cm (½-inch) EMT using one type of bender:
1. *Place the bender at the correct distance from the end of the conduit.*

   The correct distance is **equal to the desired height of the bend, minus an amount usually called the "take-up" height.** The take-up height to be applied for each bend is 12.7 cm for 1.3 cm (5 inches for ½-inch) conduit, 15.2 cm for 1.9 cm

(6 inches for ¾-inch) and 20.3 cm for 2.5 cm (8 inches for 1-inch) conduit. For the 27.9 cm (11-inch) height, subtract the take-up height of 12.7 cm (5 inches) for a remainder of 15.2 cm (6 inches).

2. *Set the mark "B" of the bender 15.2 cm (6 inches) from the end (Figure 316).*
3. *Pull on the handle with a steady pressure and press the foot step at the same time.*

Keep the conduit tight against the floor throughout the bend.

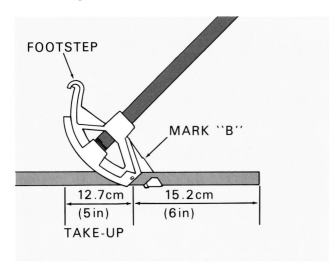

FIGURE 316. Set mark on bender as indicated.

Continue heavy foot pressure until the stub end of the pipe is at right angles to the floor (Figure 317).

NOTE: When handle of the bender is pulled to the vertical position, you have a 45° bend. For a 90° bend, continue the pressure until the stub end of the pipe is vertical and the handle is at 45° with the floor (Figure 318).

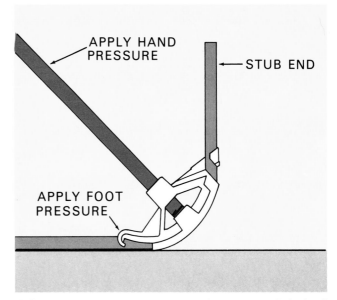

FIGURE 317. Apply hand and foot pressure until desired position is reached.

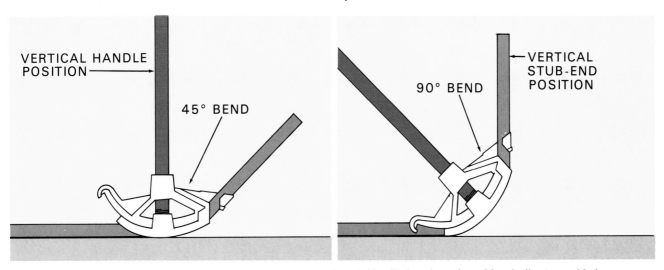

FIGURE 318. Vertical handle position indicates 45 degree bend. Vertical stub-end position indicates a 90 degree bend.

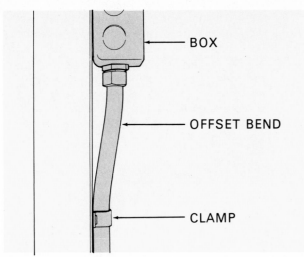

FIGURE 319. Offset bend to make straight-in connection with knockout in box.

4. *An offset bend is often required to make the conduit go straight into a box (Figure 319). Proceed as follows to make an offset bend (Figure 320):*

Draw a chalk line on the conduit when making multiple bends to guide tool placements. (Several manufacturers now provide a line on the conduit for this purpose).

Place the conduit on the floor and hook the bender close to one end. Pull on the bender until the conduit is at a **45° angle** to make the first bend. Then turn the conduit over and hook the bender under the center of the first bend. Bend again until the short end is parallel to the main piece of conduit.

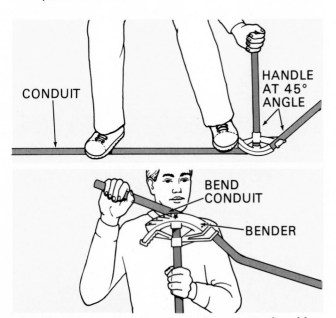

FIGURE 320. Two 45 degree bends on opposite sides produce an offset bend.

3. INSTALLING ELECTRICAL METALLIC TUBING

The conduit must be installed first and individual wires are pulled into it after it is installed. EMT must be supported within 0.92 m (3 feet) of every outlet or box and at least every 3.1 m (10 feet) between outlets or boxes.

When EMT is installed correctly, **the conduit serves as the equipment ground.** The conduit takes the place of a bare grounding wire found in NM cable, so only two wires are required inside the conduit for 120-volt or for 240-volt circuits. Circuits supplying both 120 volts and 240 volts, such as the range and dryer require 3 wires. All connections must be made very tight so they are electrically and mechanically connected at every point in the system. Only metal boxes may be used. If there is a loose connection anywhere, the conduit, boxes and equipment frames will not be correctly grounded.

EMT cannot be threaded. However, the code states that where integral couplings are used, such couplings may be factory threaded. Where coupled together by threads, the connector must be of a type to prevent bending the tubing at any part of the threads.

Two methods are commonly used to install EMT for residential wiring as follows:

1. *Drill holes through center of framing and couple short lengths of conduit for each run (Figure 321).*

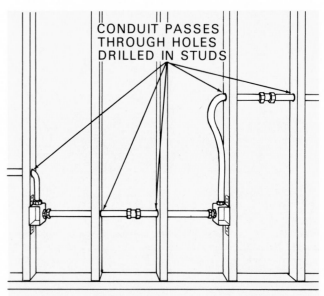

FIGURE 321. Couple short lengths of EMT through holes drilled in center of framing.

2. *Notch framing and cover notches with steel plates.*

Use this method only where necessary. Notches large enough for EMT weaken the framing to some extent (Figure 322). Use compression connectors and couplings to make box connections and to join sections of EMT during assembly.

3. *Mount the boxes loosely on the framing until the conduit connections are complete.*

If your conduit measurements are off by a fraction of an inch, you can adjust the box slightly as required. Make every connection mechanically and electrically solid to insure a good equipment ground with the conduit.

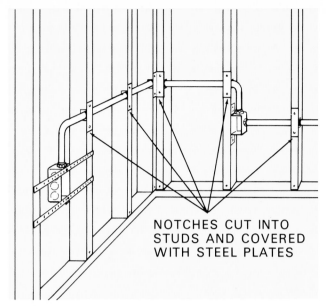

NOTCHES CUT INTO STUDS AND COVERED WITH STEEL PLATES

FIGURE 322. Notches may be used for some runs.

B. Installing Wiring in Metallic Conduit

After all conduit is installed, you are ready to pull the wiring into place. Remember that only two wires per 120-volt or 240-volt circuit are required for EMT, IMC or rigid conduit. No splices are permitted inside the conduit and no substitutions for color coding can be made. Wires must be continuous from box to box.

For short runs, you can probably push the wires from one box to another. On longer runs, or those with several bends, you need a tool called a fish-tape. The fish-tape is a flexible steel tape, usually 3.2 mm (⅛-inch) wide and 15–30 m (50- or 100-feet) long with a loop on the end (Figure 323). When pushed into the conduit, it moves more easily than insulated wires. Wires are fastened to the loop (called a hook) and pulled through the conduit. Leave 15 cm (6 inches) of wire extended for connections at each outlet.

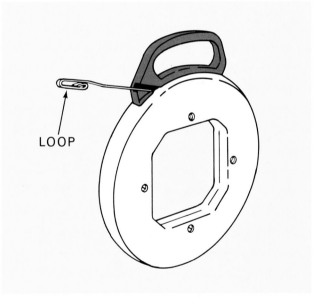

LOOP

FIGURE 323. Fish-tape helps pull wires through conduit more easily.

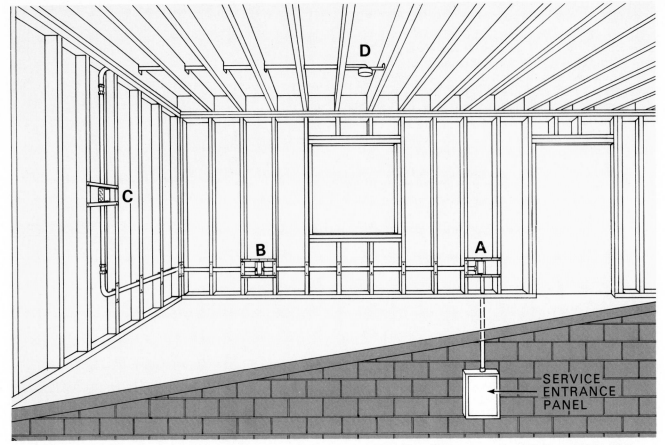

FIGURE 324. Insert wires into conduit and push or pull from point "A" to service entrance panel, from "A" to "B," from "B" to "C" and between "C" and "D."

Assume you are to pull wire into the conduit in Figure 324. Proceed as follows:

1. *Pull wires from outlet "A" to Service Entrance Panel.*

   Insert the fish-tape at the SEP in the basement and push to outlet "A" above entrance panel in the basement. Attach black and white wires to the hook. Wrap wires tightly so they won't pull loose, and make a smooth attachment so they won't catch on joints. Pull tape and wires through to the SEP. Leave enough length extended for connections at the SEP. Cut the wire at outlet "A", leaving correct length for connections.

2. *Push wires from "A" to "B" (Figure 324).*

   For a level, straight section such as this one, the tape is not needed. Push the black and white wires from "A" to "B" and cut.

3. *Pull wires between "B" and "C" (Figure 324).*

   Although short, this section contains two right angle bends, so attach black and white wires to fish tape for easier pulling.

   Insert fish tape at "C" and push to "B". Pull wires through and cut.

4. *Pull wires between "C" and "D" (Figure 324).*

   This is a switch connection and you must **use only black wires** or the color being used for hot conductors. No substitutions are allowed as in nonmetallic sheathed cable switch wiring.

   Use the fish-tape again for pushing through bends. Insert tape at "D" and push through to "C." Attach two black wires, pull through to "D" and cut.

5. *Connect wiring as for nonmetallic sheathed cable except, as noted earlier, a grounding wire is not needed.*

   The conduit itself serves as a grounding conductor.

VI. Estimating Wiring Costs

You will no doubt have had some experience as an electrician before you are called upon to estimate costs of a wiring job. Cost estimation is an important responsibility of the electrical contractor. He must be able to estimate material and labor costs accurately to stay in business.

Upon completion of this section, you will be able to follow procedures for estimating wiring costs for house wiring.

There are two ways of approaching an estimate, as follows:

A. Making a Detailed Analysis of Plans, Specifications and Labor Requirements.
B. Making a Quick Estimate.

A. Making a Detailed Analysis of Plans, Specifications and Labor Requirements

The most accurate method is to take the house plan and specifications for the job and list every detail of electrical wiring to be done. This would include a measure of the wire, conduit, or cable required. Count the number of boxes, outlets, switches and other devices needed. Include service entrance costs, plus a detailed estimate of the labor required to install all of the devices, the service entrance and wiring.

A chart similar to Tables I and II will be helpful in computing material costs.

Specifications for the wiring installation usually include the following provisions:

—The **installation** shall **conform to** NEC and local *codes.*
– **Only UL**-listed **materials** shall be used.
—**Minimum wire size** allowed in branch circuits.
—**Owner shall approve** or **select fixtures.**
—**Warranty** of materials and workmanship, usually **one year.**

After all material and labor costs have been computed, the contractor must add **overhead** costs and a **profit margin.** Overhead and profit margins may vary greatly from one section of the nation to another. Fifteen percent for each, or a total of thirty percent may be common if housing starts are high in a fast growing area. Home building may be slow in another area and the total for overhead and profit combined may be as low as 15 or 20 percent.

B. Making a Quick Estimate

There are several variations of a quick way to estimate installation costs for house wiring. Two of them are discussed as follows:

1. Averaging Outlet Costs.
2. Estimating Separate Costs.

1. AVERAGING OUTLET COSTS

In its simplest form, the contractor arrives at an average cost for installing each outlet. He then multiples the average cost by the total number of outlets in the house to arrive at an estimate of wiring costs. The total includes all receptacles, lighting outlets and switch outlets. For example, suppose a house has a total of 70 outlets. Based on past experience, the contractor figures an average cost of $15.00 per outlet for all installation costs including overhead and profit. The estimate then becomes $70 \times \$15.00 = \$1050.00$

Fixture costs for the above arrangement are arrived at by agreement on a fixed amount, usually called a fixture allowance. The amount is added to the estimate. The fixtures are selected or approved by the owner. If the cost exceeds the figure agreed on, the difference is paid by the owner.

2. ESTIMATING SEPARATE COSTS

A more accurate method is to determine separately the cost of installing 120-volt, 240-volt and 120/240-volt outlets plus the separate cost of the service entrance. Using this procedure to estimate costs for the standard house plan, proceed as follows:

NOTE: These are examples only. True costs may vary greatly, depending on local conditions.

1. *Estimate the cost for 120-volt outlets.*
   73 outlets × cost per outlet
   $73 \times \$12 = \$876.00$
2. *Estimate cost for 240-volt outlets.*
   5 outlets × cost per outlet
   $5 \times \$80 = \$400.00$
3. *Estimate cost for 120/240-volt outlets.*
   3 outlets × cost per outlet
   $3 \times \$85 = \$255.00$
4. *Estimate cost for service entrance.*
   $3 per ampere of service entrance × No. of amperes
   $\$3 \times 150 = \$450.00$
5. *Total 1 through 4.*
   $\$876 + \$400 + \$255 + \$450 = \$1981.$
6. *Add 25% for overhead and profit margin.*
   $25\% \times \$1981 = \$495$
   $\$495 + \$1981 = \$2476$
   Total estimated cost = $2,476.00

When using averages, good judgment must still be applied to each cost estimate. Look for items that might cost more than usual and include the extra costs in your estimate. For example, a 240-volt basement circuit that can run parallel with the joists costs less to install than one of the same length that must run across the joists, requiring the holes be drilled in each joist. Also, a service entrance run that is longer than usual will add considerably to installation costs.

VII. Wiring Applications for Rural Areas

If you live on a farm or in a rural area, and plan to wire one or more buildings in addition to the residence, you need to plan carefully before you begin wiring. Garages, service areas, utility buildings and barns must be included in wiring plans to provide for electrical service needs in each one.

Upon successful completion of this section, you will be able to determine the size and type of wiring needed for the rural area, to install and wire a meter pole and to install wiring in agricultural buildings.

You should consult your local power supplier for help in planning and to obtain their specific requirements. If you only have one building or service area other than the residence, you may be able to wire a circuit for it from the service entrance panel in the house, depending on the load. However, if you have two or more buildings other than the residence, you should consider installing a meter pole for use as a central distribution point to the residence and other buildings (Figure 325).

Each building must have a disconnecting means (usually called just a disconnect) and a service entrance panel. The disconnecting means may be a switch or a pullout block.

NOTE: If you have a disconnect used only for a water pump, it should be on a separate circuit ahead of the main disconnect, if it is used to provide fire protection.

To be sure of adequate power in each building, start with plans for a 3-wire service to each. The cables from the meter pole to buildings are called service drops or underground service laterals.

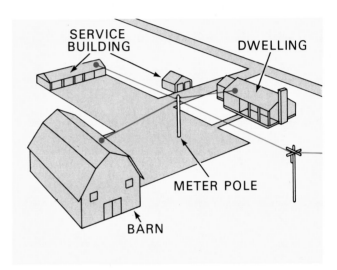

FIGURE 325. Meter pole serves as central distribution point to all buildings.

WARNING! Never attempt to disconnect power at a power supplier transformer. High voltage lines must be handled only with special equipment and by trained personnel. Many serious accidents occur annually when lay persons attempt to disconnect power at a high voltage transformer. Call your power supplier to perform this task.

Wiring applications for agricultural buildings are discussed as follows:

A. Determining Size of Wire and Service Equipment to Install.

B. Locating and Wiring the Meter Pole.

C. Wiring Service Drops to Buildings and Service Areas.

D. Wiring Agricultural and Other Utility Buildings.

E. Installing Remote-Control Wiring.

F. Installing New Cable and Boxes in Old Buildings.

A. Determining Size of Wire and Service Equipment to Install

In previous discussions of wire size as used in residential wiring, the size of wire was determined by the **ampacity** required in the circuit. In farm wiring, ampacity is also very important but two other factors must be considered. They are **mechanical strength** and **voltage drop.**

Mechnical strength is important in overhead wiring because of the possibility that wind or ice may cause the wires to break if they are not large enough. The Code requires overhead spans to be at least No. 10 wires for unsupported spans up to 15 m (50 feet) and No. 8 for longer spans.

Voltage drop is the reduction of voltage between the power source and the load, such as a motor. Voltage drop is not a problem in most residential circuits because the distances are usually not long. On farms, however, some circuits may be 92 m (300 feet) or more in length. The size of the wire must be larger on such long circuits to keep voltage drop within the normal 2 percent maximum for the load being served.

Voltage drop in wires is similar to the loss of water pressure in a long pipe. A 1.3 cm (½ inch) water pipe at 276 kPa (40 psi) would carry enough water to an irrigation outlet if it is only 4.6 to 6.1 m (15 to 20 feet) from the tank. However, if it is 153 or 183 m (500 or 600 feet) away, the pipe would be too small to maintain the desired pressure. Voltage drop in wire occurs because of total resistance on long runs if the wires are not large enough to carr the load.

Other factors shown in Table XIII affecting wire size are:
— The **type of wire** used.
— **Whether wire is run in conduit, cable, buried in the earth or overhead** in the air.
— **Whether the overhead wires are single or in triplex,** which is a 3 wire cable assembly.

Table XIII shows sizes of aluminum wire required for various amperages at distances of 15 to 92 m (50 to 300 feet), based on 2 percent voltage drop. To the left of the double line are the smallest sizes of wire allowed for the amperages shown. To the right of the double line are sizes required for various distances. For example, for a 15-ampere load, the smallest wire allowed for any length circuit is No. 12, in cable, conduit or buried in the earth, for the types of wire listed. If the circuit is overhead in the air, No. 8 wire is the smallest allowed. If the circuit were 62 m (200 feet) long, No. 6 wire would be required to carry the 15-ampere load. For a 92 m (300 feet) 15-ampere circuit, use No. 4 wire. Note that this is for aluminum conductors and for 240 volts.

Determining the size of wire and service equipment to install is discussed as follows:
1. Computing Service Load for Residences.
2. Computing Service Load for Each Building Supplied by Two or More Branch Circuits.
3. Computing the Total Load for the Entire Service.

1. COMPUTING SERVICE LOAD FOR RESIDENCES

The load for the rural residence is computed in the same way as a residence in a city. This procedure is outlined in detail in Section IV., "Installing Service Entrance Equipment."

2. COMPUTING SERVICE LOAD FOR EACH BUILDING SUPPLIED BY TWO OR MORE BRANCH CIRCUITS

If a building has two or more circuits, the load must be computed in keeping with Code Section 220–40.

Computing the service load for each building is discussed as follows:
a. Computing Lighting Load.
b. Computing Receptacle Load.
c. Computing Motor Load(s).

a. Computing Lighting Load

The lighting load should be computed at the rate of 1.5 amperes per outlet at 120 volts. Assume 10 lighting outlets in a building, for example. The total lighting load is 10 × 1.5 or 15 amperes. If a larger or smaller load is to be installed, use the actual figure, such as flood lights that might average 250 watts each. To find amperes, use the formula: amperes = watts ÷ volts.

TABLE XIII. MINIMUM WIRE SIZES FOR AMPERAGE LOADS AND DISTANCES
(Aluminum up to 400 Amperes, 230–240 Volts, Single Phase, Based on 2% Voltage Drop)

Reprinted by permission from 10th Edition, *Agricultural Wiring Handbook,* January 1993.
Used with permission of the publisher, the National Food and Energy Council, Inc., 409 Vandiver West, Suite 202, Columbia, MO 65202.

| Load in Amps | Minimum Allowable Size of Conductor | | | | | | | Length of Run in Feet | | | | | | | | | | | |
| | UF** | RH, RHW, THW, THWN, USE, NM, SE | THHN | UF** | USE | Single | Triplex | 50 | 60 | 75 | 100 | 125 | 150 | 175 | 200 | 225 | 250 | 275 | 300 |
		In Air Cable or Conduit		Direct Burial		Overhead in Air*		Compare size shown below with size shown to left of double line. Use the larger size.											
5	12	12	12	12	12	8	8	12	12	12	12	12	12	12	10	10	10	10	8
7	12	12	12	12	12	8	8	12	12	12	12	12	10	10	10	8	8	8	8
10	12	12	12	12	12	8	8	12	12	12	10	10	8	8	8	8	6	6	6
15	12	12	12	12	12	8	8	12	12	10	8	8	8	6	6	6	4	4	4
20	10	10	10	10	10	8	8	10	10	8	8	6	6	6	4	4	4	4	3
25	10	10	10	10	10	8	8	10	8	6	6	6	4	4	4	4	3	3	2
30	8	8	8	8	8	8	8	8	8	8	6	4	4	4	3	3	2	2	2
35	6	8	8	6	8	8	8	8	8	6	6	4	4	3	3	2	2	1	1
40	6	8	8	6	8	8	8	8	6	6	4	4	3	3	2	2	1	1	0
45	4	6	8	4	6	8	6	8	6	6	4	4	3	2	2	1	1	0	0
50	4	6	6	4	6	·8	6	6	6	4	4	3	2	2	1	1	0	0	00
60	3	4	6	3	4	6	4	6	6	4	3	2	2	1	0	0	00	00	00
70	2	3	4	2	3	6	4	6	4	4	3	2	1	0	0	00	00	000	000
80	1	2	3	1	2	4	3	4	4	3	2	1	0	0	00	00	000	000	4/0
90	0	2	2	0	2	4	3	4	4	3	2	1	0	00	00	000	000	4/0	4/0
100	0	1	2	0	1	4	2	4	3	2	1	0	00	00	000	000	4/0	4/0	250
115	00	0	1	00	0	3	1	4	3	2	1	0	00	000	000	4/0	4/0	250	300
130	000	00	0	000	00	2	0	3	2	1	0	00	000	000	4/0	250	250	300	300
150	4/0	000	00	4/0	000	1	00	2	2	1	00	000	000	4/0	250	250	300	300	350
175		4/0	000		4/0	0	000	2	1	0	00	000	4/0	250	300	300	350	400	400
200		250	4/0		250	00	4/0	1	0	00	000	4/0	250	300	300	350	400	400	500
225		300	250		300	000	250	1	0	00	000	4/0	250	300	350	400	500	500	500
250		350	300		350	4/0	250	0	00	000	4/0	250	300	350	400	500	500	500	600
275		500	350		500	4/0	300	0	00	000	4/0	250	300	400	400	500	500	600	600
300		500	400		500	250	350	00	00	000	250	300	350	400	500	500	600	600	700
325		600	500		600	300	400	00	000	4/0	250	300	400	500	500	600	600	700	750
350		700	500		700	300	500	00	000	4/0	300	350	400	500	600	600	700	750	800
375		700	600		700	350	500	000	000	4/0	300	350	500	500	600	700	700	800	900
400		900	700		900	400	600	000	4/0	250	300	400	500	600	600	700	750	900	900

*Single-Conductors in overhead branch circuits must be at least No. 10 copper or No. 8 aluminum for spans up to 50 feet. For overhead triplex used in free air, the smallest conductor listed in the 1990 NEC and in this Handbook edition is No. 8 AWG, even though No. 10 AWG copper has ample ampacity for small loads. For service conductors and for branch circuit spans greater than 50 feet, conductors must be at least No. 8 copper or No. 6 aluminum (NEC 225–6(a) and 230–23).

**UF not permitted in sizes greater than 4/0.

b. Computing Receptacle Load

The receptacle load should also be computed at 1.5 amperes per outlet at 120 volts. These are receptacles that are normally used for portable tools and equipment, such as drills, saws and clippers. Eight outlets, for example, at 1.5 amperes = 12 amperes total receptacle load.

c. Computing Motor Load(s)

This section must be computed in keeping with Code requirements outlined in Sections 430–22, 430–24 and Table 430–148, as follows:

—For a **single** motor, the rating must be 125 percent of the motor full-load current rating. For example, if you have only one motor in the building and its rating is 5 amperes, multiply 5 × 125% = 6.25 or 6 amperes.

—For more than one motor in a building, rate the largest motor at 125 percent and its rated full-load amperage. For additional motors and other large permanently connected equipment items, list the full amperage load.

Table XIV shows the full-load current for electric motors to help you determine ratings on motors that show only horsepower ratings.

After computing all lighting, receptacle and equipment loads for the building, add the total for all loads in the building to include lighting, receptacle and equipment loads.

NOTE: Another factor which must be considered is loads that operate without diversity, meaning, the equipment that will be operating at the same time to produce the heaviest load. The loads in each building must be considered carefully to arrive at a correct total for the load operating at the same time.

Proceed as follows to determine the load for each building supplied by two or more circuits:

1. *Add up the ampere load for all 120-volt circuits.*

   To arrive at the equivalent for a 240-volt circuit, divide the total amperage by two.

2. *Add up the amperage load for all 240-volt circuits.*

3. *Add 1 and 2 together.*

4. *As an example of a service center (shop building), list the lighting, receptacle and motor load and other large equipment as follows:*

8 lighting outlets at 1.5 a =	12 a
8 receptacle outlets at 1.5 a =	12 a
Total 120 v load =	24 a
Divide 120 v total by 2 for 240 equivalent, 24 ÷ 2 =	12 a, 240 v
Heater, 260 W =	11 a, 240 v
Tool Grinder, 373 W (½ hp) =	5 a, 240 v
Table Saw, 560 W (¾ hp) =	7 a, 240 v
Drill Press, 746 W (1 hp); 8 a × 1.25% =	10 a, 240 v
Welder =	40 a, 240 v

Total Service Center = 85 amperes, 240 volts.

NOTE: Code Table 430–148 uses headings of 115v and 230v for columns 3 and 4 - Table XIV is changed to headings of 120v and 240v to be consistent with Table XV.

TABLE XIV. FULL LOAD CURRENTS IN AMPERES FOR SINGLE PHASE AC MOTORS
(From Code Table 430–148)

(Second, third, and fourth columns from Code Table 430–148)
Reprinted with permission from NFPA 70–1993, the National Electrical Code®, Copyright 1992, National Fire Protection Association, Quincy, MA 02269. This reprinted material is not the complete and official position of the National Fire Protection Association, on the referenced subject which is represented only by the standard in its entirety.

*Power			
SI (kW)	US (HP)	120v amperes	240v amperes
.124	⅙	4.4	2.2
.186	¼	5.8	2.9
.248	⅓	7.2	3.6
.373	½	9.8	4.9
.559	¾	13.8	6.9
.746	1	16	8
1.119	1½	20	10
1.491	2	24	12
2.237	3	34	17
3.729	5	56	28
5.593	7½	80	40
7.457	10	100	50

*Table XIV provides the full load current in amperes for commercially-manufactured motors with the size designation in both SI and US units.

Add all the equipment items that are permanently connected to arrive at the load.

The loads most likely to be operating at one time to produce the heaviest loads are: the welder at 40 amperes, the heater at 11 amperes and half of the lighting at 6 amperes for a total of 57 amperes without diversity.

5. *Apply factors in Table XV to the building as follows:*

TABLE XV. METHOD FOR COMPUTING FARM LOADS FOR OTHER THAN DWELLING UNITS
(From Code Table 220–40)

Reprinted with permission from NFPA 70–1993, the National Electrical Code®, Copyright 1992, National Fire Protection Association, Quincy, MA 02269. This reprinted material is not the complete and official position of the National Fire Protection Association, on the referenced subject which is represented only by the standard in its entirety.

Ampere Load at 240 Volts	Demand Factor Percent
Loads expected to operate without diversity, but not less than 125 percent full-load current of the largest motor and not less than the first 60 amperes of load	100
Next 60 amperes of all other loads . .	50
Remainder of other load	25

a. You found that the load without diversity is 57 amperes.
b. The largest motor is one hp, drawing 8 amperes, 125% × 8 amperes = 10 amperes.
c. Not less than the first 60 amperes of all other load.

   (1) *From items a, b & c above, determine which is the larger of the three.*

   After this determination is made, apply the 100 percent demand to this amount.

   (2) *Subtract the largest of the above values, 60 amperes from the total connected load, 85 amperes, (85−60=25) amperes.*

   The next 60 amperes of the remainder has a demand factor of 50 percent. Subtract the 60 amperes from the above remainder. Since the remainder is less than 60, you show the actual amount, 25 amperes at 50 percent, or approximately 13 amperes. If the load were large enough to have a remainder after the first two steps, you would apply a demand factor of 25 percent.

   In the service center, however, the demand is 60 + 13 for a total of 73 amperes, 240 volts.

First 60 amperes at 100
   percent demand factor = 60 amperes
Remaining 25 amperes at 50
   percent demand factor = 13 amperes

Demand = 73 amperes
240 volts

Size of service entrance equipment should be 100 amperes to allow for future needs in the service center.

The wire size required for the conductors from the meter-pole to the service center is No. 1 aluminum cable in conduit or buried in the earth for distances up to 100 feet.

3. COMPUTING THE TOTAL LOAD FOR THE ENTIRE SERVICE

Repeat the process in Section 2 for each building, other than the residence.

As stated earlier, each building must have a service entrance with a disconnecting means. Size of the SEP is based on the demand load.

TABLE XVI. METHOD FOR COMPUTING TOTAL FARM LOAD
(From Code Table 220-41)

Reprinted with permission from NFPA 70–1993, the National Electrical Code® , Copyright 1992, National Fire Protection Association, Quincy, MA 02269. This reprinted material is not the complete and official position of the National Fire Protection Association on the referenced subject, which is represented only by the standard in its entirety.

Individual Loads Computed in Accordance with Table XV	Demand Factor (Percent)
Largest load	100
Second largest load	75
Third largest load	65
Remaining loads	50

To this total load, add the load of the dwelling.

To illustrate the application of Table XVI, assume you are planning a system for a farm with the following buildings and load demands, for buildings other than the residence:

Swine farrowing house	115 amperes
Service Center	73 amperes
Poultry house	36 amperes
Barn	26 amperes
Pumphouse	8 amperes

Apply demand factors from Table XVI to determine the total load. List each building with its demand factor percent.

Largest load (building with largest ampere load)
   Swine farrowing house, 115 a 100 % =   115 a
Second largest load
   Service center, 73 a at 75 % =   55 a
Third largest load
   Poultry house, 36 a at 65 % =   23 a
Remaining loads at 50 percent
   Barn, 26 a =   13 a
   Pumphouse, 8 a =   4 a
   Residence, 150 a at 100% =   150 a
   Total Farm Load =   360 a

To determine sizes of wire required for meter pole and for service drops to each building above, refer to Table XIII which is based on 2 percent voltage drop.

Suggested size of service entrance for each building is as follows:

Swine farrowing house - 115 amperes. One hundred fifty amperes recommended. This allows for future expansion if needed.

Service center - 73 amperes. One hundred amperes is the minimum size service entrance. However, the installation of one additional item of shop equipment could exceed the limit. It would be preferable to install a **150-ampere service entrance** to allow for additional load requirements.

Poultry house - 36 amperes. Sixty-ampere service entrance is recommended.

Barn - 26 amperes. The minimum size of 30 amperes is barely above present requirements. **Sixty-ampere service** entrance is recommended to allow for future needs.

Pumphouse - 8 amperes. Power requirements for the pumphouse are not likely to increase much above present needs. **Thirty-ampere service** entrance is sufficient for the single circuit.

Meter pole wires must be large enough to carry the total farm load. Based on the 360-ampere above, the meter pole wires must be 350 KCM (thousand circular mils) for cable or conduit, Types UF, RH, THW, USE, RHW, THWN, NM, SE, THHN, at 38.1 m (125 ft.). Instructions for wiring meter poles are given in the following section (B).

Wire sizes for each building above are as follows, as indicated in Table XIII for Types R, T and TW at various distances.

Swine farrowing house - 150 amperes, No. 2/0 for distances to 30.5 m (100 feet) and No. 3/0 45.8 m (to 150 feet).

Service center - 150 amperes, No. 2/0 for distances to 30.5 m (100 feet) and 3/0 to 45.8 m (150 feet).

Poultry house - 50 amperes, No. 4 AWG wire size for distances to 30.5 m (100 feet) or No. 2 to 45.8 m (150 feet).

Barn, 60 amperes, No. 3 wire for distances to 30.5 m (100 feet) or No. 2 to 45.8 m (150 feet).

Pumphouse - 30 amperes, No. 6 wire for distances to 30.5 m (100 feet) or No. 4 for 45.8 m (150 feet).

Residence - 150 ampere, 2/0 wire for distances to 30.5 m (100 feet) and 3/0 for 45.8 m (150 feet).

Motor controllers (switches) and terminals of control circuits must be connected with copper conductors only.

B. Locating and Wiring the Meter Pole

Locating and wiring the meter pole is discussed as follows:

1. Locating the Meter Pole.
2. Wiring the Meter Pole.

1. LOCATING THE METER POLE

One reason for installing the meter on a pole is to reduce the length of wire or cable required for two or more buildings. Other advantages are that repairs or changes can be made to wiring in one building without affecting service to the others and, usually, smaller conductors can be installed to service each building because of shorter distance.

Normally, the meter pole is located near a point in the center of the electrical load to be served. An exception might be if load demand to one building is much greater than the others. In such a case, it may cost less to locate the meter pole somewhat nearer the heavy demand to reduce the amount of larger wire required. Select the location that is most convenient and least costly (Figure 326).

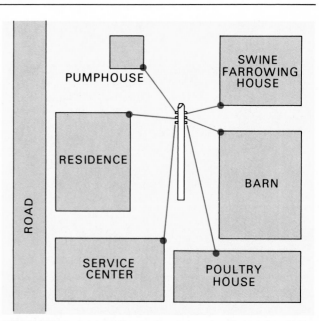

FIGURE 326. Locate the meter pole near the center of the electrical load to be served.

2. WIRING THE METER POLE

In some localities, the power supplier installs the meter pole. The power supplier also brings the wiring to meet the owner's wiring at the top of the pole. It is usually the owner's responsibility to install the wiring from that point. In every case, the power supplier should be contacted for specific requirements for the installation. In many instances, **current transformer** metering is installed on the meter pole, especially where the service size exceeds 200 amperes. Where current transformer metering is installed, large conductors are not needed up and down the pole. You must know the requirements of your local code and power supplier before you begin. In all cases, the neutral is grounded at the meter pole, using a different procedure than for grounding house wiring, as explained in the next section.

Wiring the meter pole is discussed as follows:
a. Wiring the meter pole with one conduit.
b. Wiring the meter pole with two conduits.
c. Wiring the meter pole with service entrance cable.
d. Wiring the meter pole when current transformer (CT) metering is used.

a. Wiring the Meter Pole with One Conduit

NOTE: The conduit with wires installed, the meter base, the disconnect switch and ground wire should be mounted on the pole before the pole is erected.

In this installation, you will run the two hot wires and the neutral wire for the power line connection at the top of the pole, down to the meter socket. Only the two hot wires run back to the top of the pole.

Proceed as follows:
1. *Install insulator racks or eyebolts and connectors near top of pole.*

Mount one rack or connector for conductors from the power supplier, and one for each building that will be served from the pole (Figure 327). Fasten with heavy screws, or with bolts through the pole if heavy icing conditions are likely.

2. *Install the conduit and entrance head.*

Select conduit type and size as in Section V and install on pole. Mount the top of the conduit higher than the insulators to provide for a drip loop. Fasten conduit in place with conduit straps. Attach entrance head with required number of holes (Figure 328).

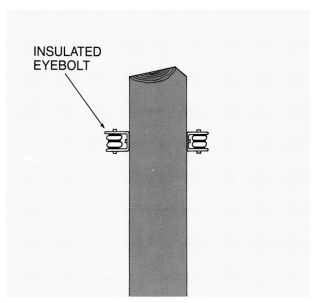

FIGURE 327. Mount one rack or eyebolt for power supplier conductors and one for each building to be served.

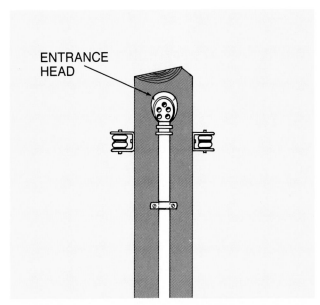

FIGURE 328. Select an entrance head with the required number of holes and mount high enough to form drip loops.

3. *Install meter base.*

Follow procedures as in Section D. Mount the meter about 1.4–1.7 m (4½–5 feet) above ground (Figure 329).

4. *Install wiring.*

Push the neutral wire (white) and two hot wires (usually black) down through the conduit to the meter base.

Leave about 1.2 m (48 inches) at the top of the pole to be attached to the power supplier lines. Connect the neutral and two hot wires to the meter base.

> **NOTE: The Code does not allow a disconnect to be installed where only one conduit is installed for meter pole wiring.**

5. *Install hot wires from the meter base to the distribution wires.*

Inside the same conduit, install two more hot wires. The neutral connections are made at the top of the pole by splicing each one to the power line neutral, so another neutral is not required in the conduit.

Connect the bottom end of the two hot wires to the lower (output) terminals on the meter base. At the entrance head, you will connect the two hot wires to the outgoing wires to other buildings when installed (Figure 330).

6. *Connect neutral from power line.*

Splice a wire between the power line neutral and the outgoing circuit neutrals, using a heavy split-bolt connector. Although split-bolt connectors are available this type of connection is not the preferred connector for most utilities or contractors. A compression type fitting is far superior over the life of the installation. Illustrations 330, 331, 332 are shown with split bolt connectors; while this type termination is acceptable a compression type connection would be far superior (Figure 331).

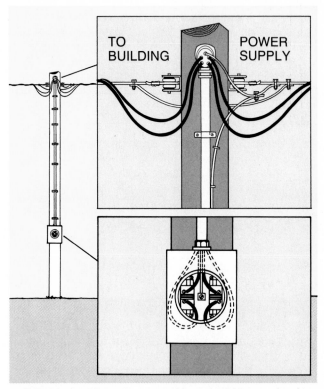

FIGURE 330. Wiring in meter base and at top of pole.

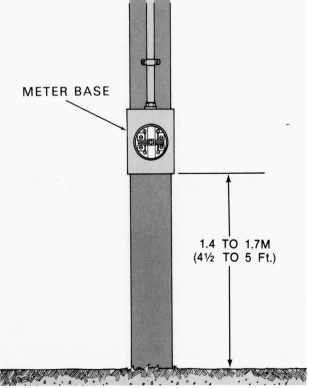

FIGURE 329. Mount the meter base 1.4 to 1.7 m (4½ to 5 feet) above ground.

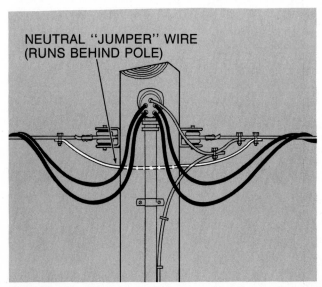

FIGURE 331. Wire between power supply neutral and outgoing neutrals.

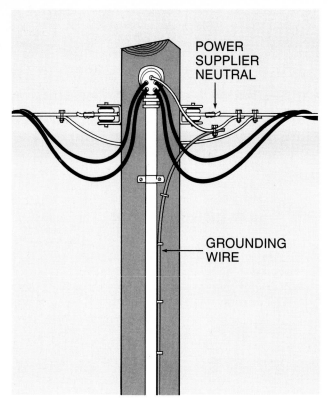

FIGURE 332. Grounding wire connected to power supplier neutral and stapled to pole.

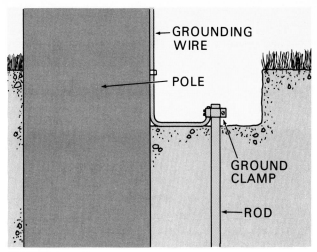

FIGURE 333. Grounding wire attached to grounding rod with approved clamp.

7. *Connect the grounding wire.*

On previous ground connections, you have connected the grounding wire to the service entrance and from there to the cold water pipe and/or a "made" ground. Because of the increased danger from lightning on pole installations, most power suppliers recommend connecting the grounding wire to the neutral at the top of the pole.

Splice the grounding wire to the neutral and run it down the pole direct to the ground rod or water pipe (Figure 332). No. 6 wire is usually large enough, but some power suppliers require No. 4. Staple the grounding wire to the pole alongside the conduit for protection, or to the opposite side of the pole, if your Code requires it. To protect the grounding wire when installed with a ground rod, dig a hole 15 or 20 cm (6 or 8 inches) deep next to the pole. Select and install the rod as in Section IV, "Installing Service Entrance Equipment." A 1.3 cm (one half inch) copper rod 2.4 m (8 feet) long is a common size. Attach the ground wire with an approved clamp and cover with soil after inspection (Figure 333). If the ground wire attached to the pole is subject to damage by vehicles driving past or other hazards, nail a narrow strip of wood to cover it to a height of 2.4 to 3.1 m (8 to 10 feet), for protection.

All cables must be secured within 8 inches of each cabinet, box or fitting.

NOTE: Grounding and bonding of agricultural buildings must comply with the exceptions noted in article 547-8 Ex (1)(b) of the Code. This exception is especially important since it provides for a full size equipment grounding conductor in Agricultural buildings. Providing this full size ground will tend to reduce the amount of voltage buildup on the equipotential grid (grounding electrode system) during a fault condition. Also a full size equipment grounding conductor will reduce the step potential or gradient voltages caused by leakage currents that are present in all systems. These voltages cause havoc in agricultural buildings.

b. Wiring the Meter Pole with Two Conduits

Proceed as follows:

1. *Run the neutral and two hot wires in one conduit for connecting the power supplier lines to the meter base (and to a disconnect switch, if installed).*

2. *Mount a second conduit with neutral and two hot wires to run from the meter base or disconnect switch to the top of the pole (Figure 334).*

Since circuits in both conduits have neutral wires, it is not necessary to connect the neutral from the power line to neutrals of outgoing circuits. Instead, connect the neutral from the meter base or disconnect switch to the outgoing circuits at the top of the pole.

Check with the utility to determine if a disconnecting means is required for this installation. Most utilities will prefer that a disconnect be installed for safety reason. If this switch is designed to interrupt only the commercial power supplied by the utility then a double pole single throw will suffice. However if a emergency generator is installed

and this disconnect must serve an additional function then a double pole double throw switch must be installed. The additional throw of the switch will first disconnect the commercial power from the additional buildings served and connect the output of the emergency generator to the additional buildings previously disconnected. This isolation or transfer switch is required by code 230-83 to first separate completely from commercial power before connecting to the generator output. Failure to isolate the two systems completely can cause loss of life and property, including damage to the generator and associated equipment. Essential to the safe operation of this transfer switch is a neutral position where neither the commercial power nor the emergency power is connected to the load. This neutral position will allow the generator to attain proper running speed and output voltage prior to the generator output being connected to the load. Connecting the generator to a load before it is allowed to attain running speed or proper utilization voltage can cause severe generator damage and damage to utilization loads.

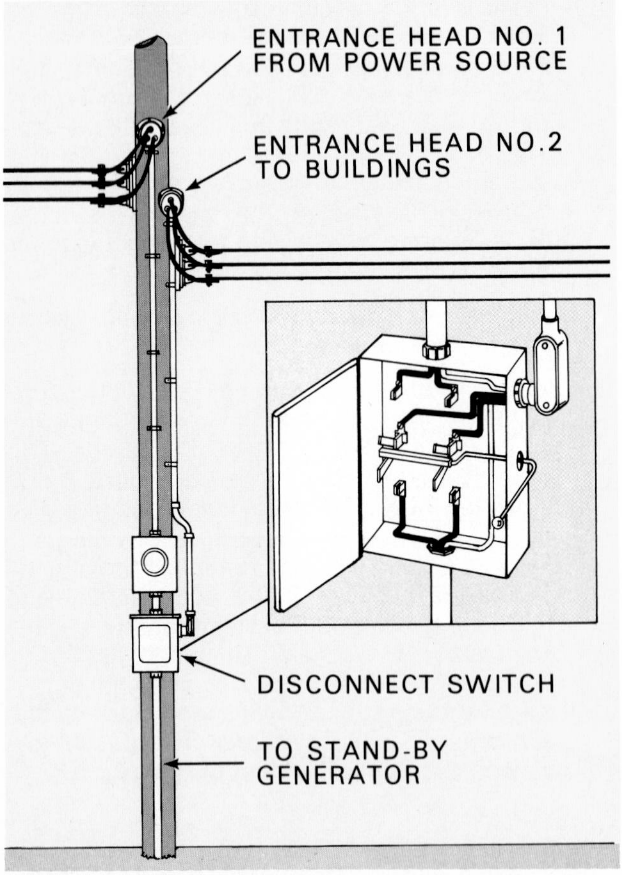

c. Wiring the Meter Pole with Service Entrance Cable

Install three-wire service entrance cable following the same procedures outlined above for two conduits. Fasten the cable with straps and protect it from damage where required. This switch and service must be grounded to a grounding electrode as shown in figure 333.

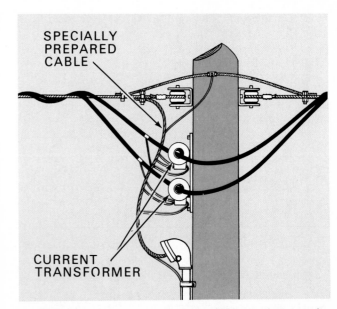

FIGURE 335. Current transformer (CT) metering permits the use of smaller gage wiring from power supplier wires to the meter.

ENTRANCE HEAD NO. 1 FROM POWER SOURCE

ENTRANCE HEAD NO. 2 TO BUILDINGS

DISCONNECT SWITCH

TO STAND-BY GENERATOR

FIGURE 334. Two conduits each contain neutral and two hot wires. They connect power supplier lines to meter base and meter base to outgoing wires at the top of the pole.

168

d. Wiring the Meter Pole When Current Transformer (CT) Metering Is Used

Current transformer metering is now frequently installed on the meter pole, especially where total loads approach or exceed 200 amperes. In these installations, large conductors are not needed down and up the meter pole (Figure 335).

The current transformer is normally installed on the meter pole by the power supplier. Operating at a maximum of 5 amperes to the meter, small gage wires are installed between the current transformer and meter. Service drops to buildings are spliced directly to the service drop conductor from the power supplier unless there is a pole-top switch installed.

C. Wiring Service Drops to Buildings and Service Areas

Refer to Table XIII to determine the size of wire needed for each service drop or underground service lateral.

Wiring service drops to buildings is discussed as follows:

1. Wiring Overhead Service Drops to Buildings.
2. Wiring Underground Service Laterals to Homes and Buildings.

1. WIRING OVERHEAD SERVICE DROPS TO BUILDINGS

1. *Cut conductors to length.*

   Unroll and cut the conductors (triplex or individual wires) to required length on the ground. Allow about 1.2 m (four feet) at each end for attaching to insulators.

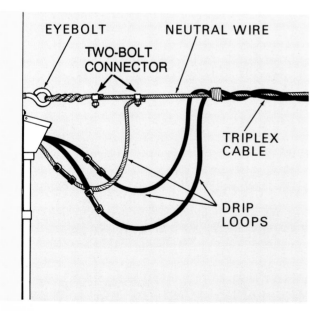

FIGURE 336. Triplex cable is usually fastened to the building by means of an eyebolt or insulated connector.

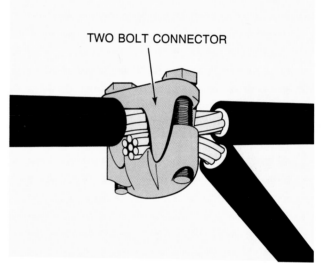

FIGURE 337. Two-bolt solderless connectors are used to splice the heavy outdoor wires. Splice must be wrapped with insulating tape.

2. *Attach conductors to insulator(s) on pole and to building.*

   Triplex cable is generally favored for service drops. It is made up of a bare neutral wire with two insulated cables wrapped around it for support. Triplex cable is usually fastened to the meter pole and to the building by means of an eyebolt or insulated connector (Figure 336). Fasten the eyebolts securely to the pole and buildings. If mounting on a wood wall, screw the eyebolt directly into a stud or other framing member in back of the siding or sheathing. Loop the neutral wire through the eyebolt and fasten with a crimp fastener or two-bolt connector. There are also wedge-type devices for holding and securing the bare messenger of the triplex.

3. *Splice wires for feeder connections.*

   Splice the heavy outdoor wires with two-bolt aluminum solderless connectors. They may be used to tap a wire, as with drip loop connections (Figure 337), or to splice two ends together. Proceed as follows:

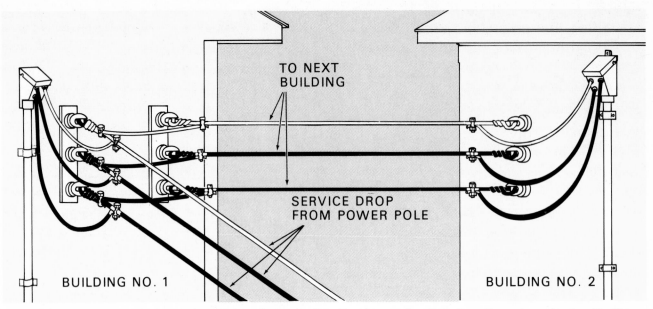

TO NEXT BUILDING

SERVICE DROP FROM POWER POLE

BUILDING NO. 1

BUILDING NO. 2

FIGURE 338. Wiring for two buildings fairly close together. Tap the service drop from the pole and extend to the second building.

Remove the insulation from the wires and place together. Remove the two-bolts from the connector. Apply inhibitor paste to the wires and connector. Place the connector bolts over the wires. Replace the bolts and tighten with a wrench. Cover the splice with insulating tape. Wrap the tape around the splice to a thickness equal to conductor insulation or greater.

NEC® Code 230-9 requirements state that service conductors must not be installed beneath openings through which materials may be moved. An example is a hayloft or a barn. A bale of hay dropped on a service cable below could pull it off its mounting. Overhead wire must not obstruct the entrance to a door.

4. *Attach wires to house or other buildings.*

Install the insulator or insulator rack as in Section IV. C., "Locating and Mounting the Service Drop." Procedures are the same. Remember to observe the height requirements for the overhead wires as in Section IV. C.

5. *Install service entrance.*

Follow procedures in Section IV. D., "Installing Service Entrance." Procedures are the same except that you do not install a meter.

A 30-ampere capacity service entrance may be installed if the building does not have over 2 circuits. For all others, install service equipment rated according to directions in Table XV. Remember that each service entrance must have a disconnect, overload protection and must be grounded.

If you have two buildings close together and both contain fairly small loads, you may serve both with one circuit from the pole. Use wires large enough to carry both loads. Run the three wires from the pole to one building. Then extend wires to the next building as shown in Figure 338. If only 120 volts are needed in the second building, tap only the neutral and one hot wire for a 2-wire circuit. Install complete service entrance at each building.

It is permissible for service conductors to be installed on the exterior of one building to serve a second building. The service conductors must not pass through the interior of one building to serve another unless both buildings are under one management or occupancy. A grounding electrode is not required at the second building if it contains only one branch circuit.

2. WIRING UNDERGROUND SERVICE LATERALS TO HOMES AND BUILDINGS

Underground wiring is becoming popular for use in homes, farms, and many other places. It is simple to install with trenching equipment. Newer types of wire are low in cost and there are fewer maintenance problems. With underground wiring, there are no service interruptions caused by ice storms, falling branches and high winds. Tall equipment can be moved freely without the hazard of striking overhead wires.

Two types of cable are approved for underground wiring. They are Type UF (underground feeder) and Type USE (underground service entrance). Both types can be used to run feeder circuits buried directly in the earth from the pole to homes and buildings. Special installation precaution preclude either of these cables being

170

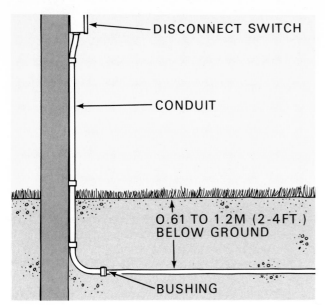

FIGURE 339. Conduit should extend 0.61 to 1.2 m (2 to 4 feet) below ground with bushing attached to protect cable.

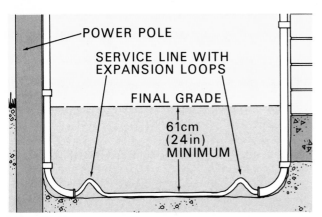

FIGURE 340. Lay cable in a trench at least 61 cm (24 inches) deep. Leave expansion loop at each end.

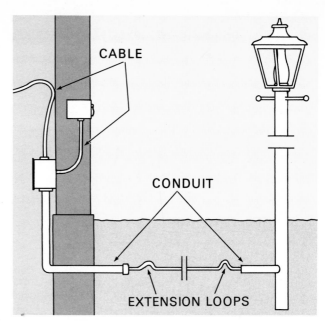

FIGURE 341A. Circuit extension for post lantern.

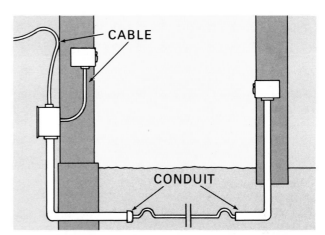

FIGURE 341B. Circuit extension to another building.

installed; type UF cable where run from a pole must have fuses or circuit breakers installed at it origin(pole); USE conductors can be used as service conductors when it is protected at its termination; UF type cable must be used as a feeder which by its name implies that it is protected where it receives its voltage supply.

Proceed as follows:

1. *Install conduit with wire or service entrance cable from top of pole to meter base and to the disconnect switch.*

   Leave 1.2 or 1.5 meters (four or five feet) at top of pole for connections to power suppliers service drop.

2. *Install conduit from disconnect switch to trench for underground cable.*

   Conduit end at trench must have bushing installed to prevent damage to cable from rough edges (Figure 339). Conduit should protect cable from the switch to a point 0.61 to 1.2 m (2 to 4

feet) below grade level.

3. *Install cable in trench.*

   Trench must be at least 61 cm (24 inches) deep. Local codes may require up to 1.2 m (48 inches). Lay cable with expansion loop at each end (Figure 340). When cable is located beneath driveways or other heavily loaded surfaces or if sharp rocks may damage the cable, protection should be provided with conduit. Underground conductors run under a building must be in conduit. The conduit must extend beyond the outside walls of the building. The Code doesn't say how far.

   Similar procedures may be followed to extend a circuit from inside the house to a post lantern (Figure 341A) or to another buildings (Figure 341B).

4. Follow procedures in Section IV. D4., "Installing Underground Service Entrance Cable."

171

D. Wiring Agricultural and Other Utility Buildings

Wiring agricultural buildings is discussed under the following headings:
1. Types of Wire.
2. Types of Wiring Devices, Boxes and Fittings.
3. Procedures for Wiring Buildings.
4. Stray Voltage.

1. TYPES OF WIRE

Type NM nonmetallic sheathed cable can be used only in normally dry locations. For wiring most agricultural buildings, either Type NMC, which may not be available or Type UF (underground feeder) cable is recommended. Both have insulation and covering that permits their use in dry, moist, damp or corrosive locations and in outside or inside walls of masonry, block or tile. An added advantage for Type UF is that it may be used underground. Also, it may be exposed to sunlight on the exterior of a building if not subject to physical damage.

All cables should be surface mounted to permit proper wiring inspection and maintenance, and to minimize damage from rodents. The cables should be secured by non-metallic cable straps with corrosion resistant nails. The cables should be secured at three-foot intervals on vertical surfaces and two-foot intervals on horizontal surfaces and within eight inches of each box or device.

Portable tools and equipment are likely to be used at times in most agricultural buildings. To insure equipment grounding at every outlet, install cable and receptacles with a copper equipment grounding conductor required by *NEC*® 547-8(c) in all agricultural buildings.

Metal conduit and armored cable should not be installed in agricultural buildings where moist and/or corrosive conditions exist. Rust and corrosion may gradually cause separation of the joints. This action exposes conductors inside the conduit and may cause a short circuit. Also, it leaves the circuit without effective equipment grounding since conduit or armor serves as the equipment grounding conductor.

If more than 3 conductors are needed, as for a 3 phase fan motor or control, then it is convenient to use **non-metallic conduit**. There are several types of conduit available from different manufacturers. Some insurance companies are concerned about toxic fumes from this material (in instances of fire) so it is advisable to check with your insurance company before installing.

In agricultural buildings, conduit may be subject to severe physical abuse so heavy walled conduit should be used. Type 80 heavy-walled rigid PVC conduit (Figure 342) is designed for both above ground and underground applications. It should not extend from a warm, moist part of a building to a cold part because of condensation collecting in the conduit. **Liquidtight non-metallic flexible conduit** has been developed to replace electrical metallic tubing (EMT) (thin-walled conduit). Several types have concentric corrugations that give them high strength and bendability without heavy weight (Figure 342). Special couplings and quick connect terminations fit into the concentric corrugations to make it easy to attach to special single-gang or two-gang non-metallic boxes. This type of watertight non-metallic conduit is used in such wet locations as milking parlors and fruit and vegetable processing areas.

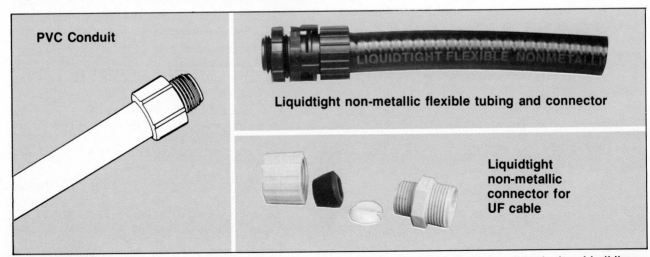

PVC Conduit

Liquidtight non-metallic flexible tubing and connector

Liquidtight non-metallic connector for UF cable

FIGURE 342. Non-metallic surface-wiring devices are often installed in Group I and Group II agricultural buildings.

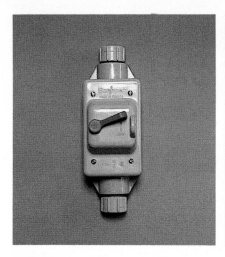

FIGURE 343. Group II agricultural buildings require devices, boxes and fittings that are both dust-tight and watertight and made of corrosion-resistant materials.

2. TYPES OF WIRING DEVICES, BOXES AND FITTINGS

The Code restrictions for agricultural buildings, Article 547, make it desirable to divide farm buildings into two types for selection of wiring devices, boxes and fittings. For purposes of this discussion, they will be divided into Group I and Group II.

Included in Group I agricultural buildings are those that do not require special materials for wiring. Examples are open barns, service centers and machinery and equipment storage buildings. Care should be exercised when selecting an appropriate wiring method for this group of agricultural buildings. This type structures has condensation and all types of dust. It's clear that the adverse conditions found in some agricultural buildings does not exist here, thus some less restrictive methods may be employed. Conditions of the most adverse environments may require the wiring methods described in *NEC®* 547-1(a) are acceptable. Where severe conditions do not exist the wiring methods other than those described in *NEC®* 547 are acceptable.

Group I buildings may be wired with the same type of materials used for residential wiring in most cases. **Number 12 AWG should be the minimum size wire used in agricultural buildings.** Surface wiring is often installed in farm buildings to save on materials and labor and prevent damage from rodents. You do not drill holes through framing members for surface wiring. Switches, duplex receptacles and lighting outlets may be self-contained.

Group II buildings are those that house livestock or poultry in confinement and are totally enclosed and environmentally controlled or in buildings or areas of buildings where one or more of the following conditions exist:

—Where **excessive dust** or **dust with water** may accumulate from litter dust or feed dust.
—Where **poultry and animal excrement** may cause corrosion vapors in the confinement area.
—Where **corrosive particles** may combine with water.
—Where the **area is damp and wet** by reason of periodic washing with water and sanitizing agents.

Where required, insurance rates may be reduced for those who install approved wiring and materials.

Such buildings must have materials installed that meet strict Code requirements, as follows:

—Cables shall be Types UF, NMC, SNM, Copper SE or other cables or raceways suitable for the location.
—All wiring devices, boxes and fittings must be dust-tight and watertight and made of corrosion-resistant materials (Figure 343).

Lighting fixtures in agricultural buildings must comply with the following:

—A lighting fixture that may be exposed to moisture from condensation or building cleaning water must be watertight.
—Any lighting fixture that may be exposed to physical damage must be protected by a suitable guard (Figure 344). Lighting fixtures must be installed to minimize entrance of dust, foreign matter, moisture and corrosive material.

Switches, **circuit breakers**, **motor controls** and **fuses** must have a weatherproof, corrosion-resistant enclosure with a close-fitting cover. Nonmetallic boxes are available that are corrosion resistant and very durable. Such boxes may be fitted with dust-tight and watertight devices. Also, complete units are available that will meet the requirements (Figure 345).

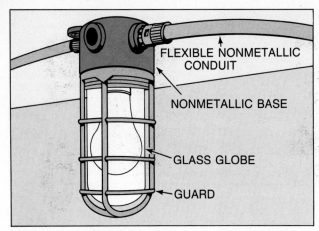

FLEXIBLE NONMETALLIC
CONDUIT

NONMETALLIC BASE

GLASS GLOBE

GUARD

FIGURE 344. Any lighting fixtures in Group II buildings that may be exposed to physical damage must be protected by a suitable guard.

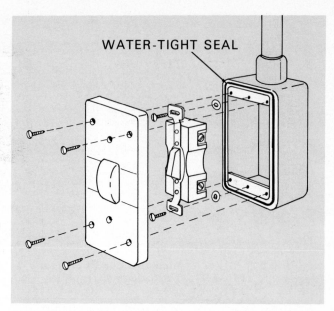

WATER-TIGHT SEAL

FIGURE 345. Nonmetallic boxes and corrosion-resistant, dust-tight and watertight devices are required for Group II buildings.

In Group II buildings, non-metallic conduit or UF cable is the preferred wiring material. All cable, cord, and conduit must have an equipment grounding conductor. Where flexibility is needed, cord with appropriate fittings or liquidtight flexible non-metallic conduit can be used. The combination of non-metallic conduit and UF cable with matching fittings can provide a corrosion resistant watertight wiring system which will withstand the harsh environment of Group II buildings. Where conduit is used, it should be sealed and not run through attic spaces.

Motors and other rotating electrica machinery must be totally enclosed or otherwise fully protected from dust, water and corrosive particles *NEC*® 547-6. Grounding must comply with instructions in Code article 547-8(a).

Exception No. 1 requires that the equipment grounding conductor in a cable must have an ampacity equivalent to the circuit conductors.

Exception No. 2 permits the bonding of metal water piping or other metal pipes to an impedance device listed for the purpose. This device is intended to eliminate neutral voltages on interior piping systems.

Following are other items insurance companies recommend producers give consideration when building Group II livestock or poultry buildings:

—Install as much as possible of the electrical system, such as the SEP, time clocks, and switches, outside the animal area in an office or workroom. If the SEP is located in the livestock area, it should be mounted in a moisture resistive enclosure, on fire-resistive material such as asbestos board.

—The SEP should be located on an interior division wall.

—In those areas where the wiring system is subject to physical damage (such as inside a pen of livestock), the electrical wire should have physical protection from damage such as with the use of schedule 80 rigid PVC conduit. A separate grounding wire must be installed in all cable conduit in such areas as required by the Code. This would insure grounding continuity in case the conduit separates at some point.

—If the service entrance head is located outside of the building, the service entrance conduit should be connected to either the side or bottom of the SEP.

— Electrically-heated livestock waterers or water lines heated with an electric heat tape should be grounded utilizing a two wire w/g system for protection from electrocution.

—A lightning surge arrestor should be installed on the building's incoming electrical service entrance lines or at the SEP.

—The electrical power supplier should be requested to review the electrical blueprints for the building and/or to visually inspect the electrical apparatus for an auxiliary powered generator should one be utilized.

—The electrical circuitry for the water supply system of the building should be wired ahead of the main electrical disconnect at the building. Then if the electrical system at the building is shut off, there would still be a supply of water to the structure.

—All well pumps are required to be grounded unless specifically designed to be fully insulated from ground. Where submersible pumps are used in metallic well casings, the well casing must be bonded to the pump circuit equipment grounding conductor or the SEP.

174

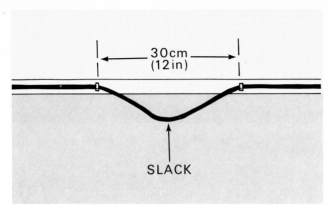

FIGURE 346. Surface-mounted devices require only short lengths of wire for terminal connections. Leave 30 cm (12 inches) of slack at each outlet.

3. PROCEDURES FOR WIRING BUILDINGS

Wire Group I agricultural buildings following similar procedures as for residential wiring except that the cable is mounted on the surface. The special surface-wiring devices that may be used include duplex receptacles, lampholders, single-pole switches and three- and four-way switches. They are shock-proof and corrosion-resistant. Some types provide extra terminals, allowing additional circuits to be added later as needed.

Proceed as follows to wire surface-mounted circuits and devices.

1. *Run cable on surface from SEP to all switches and outlets on circuit.*

Leave 30 cm (12 inches) of slack for each device to be mounted (Figure 346). Terminals in surface mounted devices eliminate the need for splices and solderless connectors for some types. Therefore, only short lengths of wire are needed for terminal connections at each outlet. Run cable in protected areas of joists, timbers and studs where possible, following closely the contours of the building (Figure 347).

2. *Mount surface-type switches, lampholders and receptacles.*

Mount overhead lampholders in recessed positions between joists for protection. A 2.5 cm × 10.2 cm (1″ × 4″) or 5.1 cm × 10.2 cm (2″ × 4″) nailed between or across top edges of joists permits lampholders to be mounted directly on the surface (Figure 347).

To mount a device consisting of a plastic outlet box with threaded knockout, a ½″ plastic threaded cable connector and plastic keyless lampholder, proceed as follows:

(1) *Remove knockouts as needed and mount the outlet box using two nails or screws.*

(2) *Insert the threaded cable connector into the knockout opening and tighten (Figure 348).*

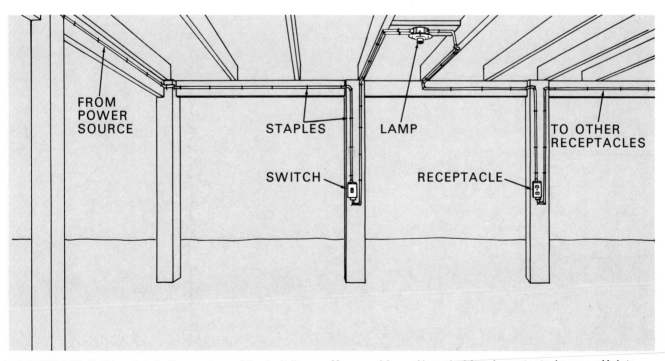

FIGURE 347. Follow closely the contour of the building and keep cable and lampholders in protected areas of joists, timbers and studs.

175

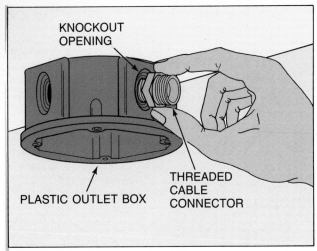

FIGURE 348. Insert the threaded connector into the knockout opening.

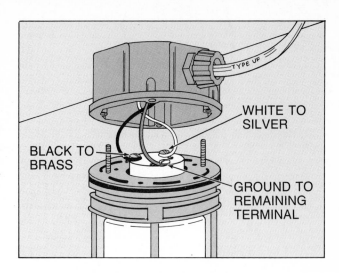

FIGURE 350. Connect the wires to the terminals—white to silver, black to brass, and grounding wire to ground terminal.

(3) *Remove insulation from wires and slide the connector cap, plastic clamp and rubber gasket over the wires (Figure 349a).*

(4) *Insert the wires into the knockout opening.*

(5) *Slide the rubber gasket, clamp, and cap against the cable connector and tighten the threaded cap (Figure 349b).*

(6) *Pull the wires downward and connect them to the lampholder screw terminals—white to silver, black to brass and ground wire to remaining terminal (Figure 350).*

(7) *Fold wires into the outlet box and attach lampholder to the outlet box (Figure 351).*

Install a gasket between the outlet box and lampholder if one is available.

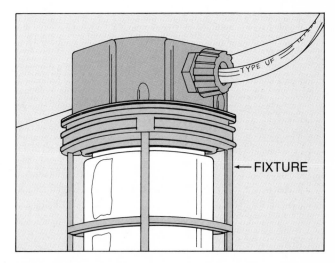

FIGURE 351. Fold the wires into the outlet box and attach the fixture.

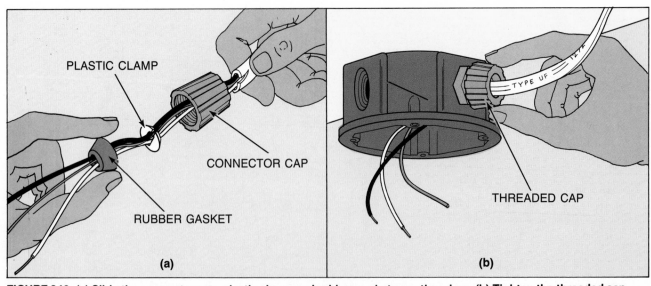

FIGURE 349. (a) Slide the connector cap, plastic clamp and rubber gasket over the wires, (b) Tighten the threaded cap.

176

4. STRAY VOLTAGE AND THE EQUIPOTENTIAL PLANE

Small electrical currents passing through livestock in confinement situations can cause sever production losses and poor animal health. These small electrical currents are usually caused by low alternating current (AC) voltages on the grounded neutral system and also in the equipment grounding conductor on the farm wiring system and are often referred to as **stray voltage**. The voltage and resulting currents may originate off the farm and be brought to the farm through the grounded neutral wiring network.

Authorities estimate that 20 percent or more of the nations dairies have stray voltage in the milking parlor exceeding ½-volt (AC). Voltages of alternating current exceeding ½-volt between feed troughs or other metal to wet concrete floors can be detected by livestock. The resulting "shock" and aggravation can lead to substantial monetary losses. One case study documented that an 80-cow dairy herd suffered a 3,000 pound reduction in milk production per cow over a period of two years, and over $15,000 was spent in an attempt to solve the problem.

Reduction of stray voltage in livestock confinement areas generally can be accomplished if a careful analysis of the problem is made. Proper measurement and application of recommended techniques in construction and bonding provide the most satisfactory approach for eliminating stray voltage. An **equipotential plane** may be used to help control stray voltage. The equipotential plane consists of an area where a wire mesh or other conductor is embedded in concrete, bonded to all adjacent conductive equipment, structures or surfaces and connected to the electrical grounding system. Its purpose is to prevent a difference in voltage from developing within the plane. With this arrangement, livestock making contact between the floor and the equipment will be less likely to be exposed to a level of voltage that could cause problems.

Wire mesh or other conductive material installed to provide an equipotential plane must be bonded with a copper conductor not smaller than No. 8. Pressure connectors or clamps of brass, copper, or copper alloy must be used to make the bond.

In Group II agricultural buildings, non-current carrying metal parts of equipment, conduit, and other enclosures required to be grounded, must be grounded by a copper equipment grounding conductor. This conductor must be installed between the equipment and the building disconnect. If installed underground, the conductor must be insulated or covered.

To obtain more information on the symptoms, causes, measurement and reduction of stray voltage, contact your state land-grant university or your local cooperative extension service office.

Sources of additional information on agricultural building wiring and electrical systems are listed in the appendix of this publication.

E. Installing Remote-Control Wiring

Remote-control low-voltage (24-volt) systems offer some advantages for use in agricultural wiring to control lights from several locations. Also, some people prefer to install remote-control switching systems in residences.

The outlets for remote-control systems, such as lights, are wired in the usual way with 120-volt circuits. The advantage lies in the low-voltage, inexpensive cable used for the control circuits to the switches. Number 18 insulated cable may be used. If you have a central yard light, for example, that you want to control from two or three locations that are a considerable distance apart, low-voltage wiring to each switch is less expensive. And because it is low-voltage, it is safer.

The remote-control system consists of the following materials other than those usually installed (Figure 352): **one low voltage transformer 24 volts, No. 18 three conductor cable for outdoor use, a low-voltage remote-control relay at each outlet, and push-button switches.** Also available is a weather-resistant metal outdoor unit containing a built-in 24-volt transformer and control relay (Figure 353). The unit has two compartments, the upper one for low-voltage connections and the lower for 125 volts. The unit is mounted with two machine screws. A circuit wiring diagram is included

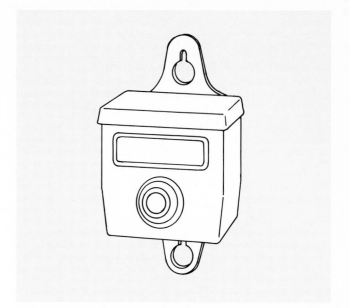

FIGURE 353. Weather-resistant outdoor unit.

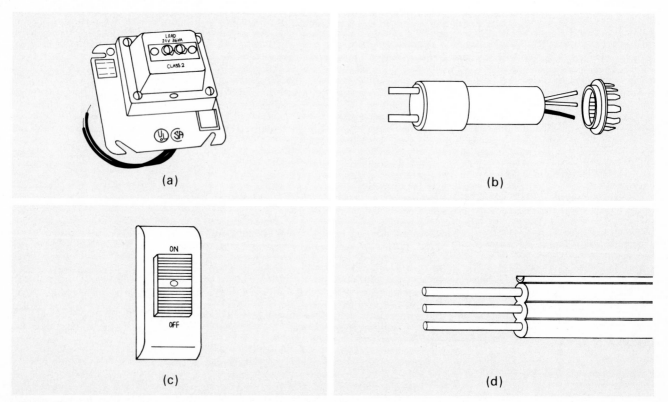

(a)

(b)

(c)

(d)

FIGURE 352. Components of the remote-control system are (a) Low-voltage transformer, (b) Low-voltage remote-control relay, (c) Low-voltage switch and (d) Three-conductor low-voltage cable.

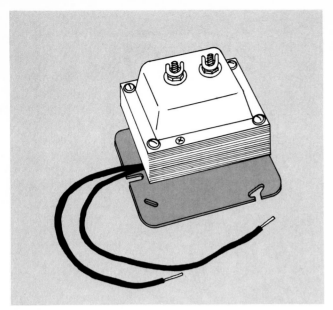

FIGURE 354. Twenty four-volt transformer.

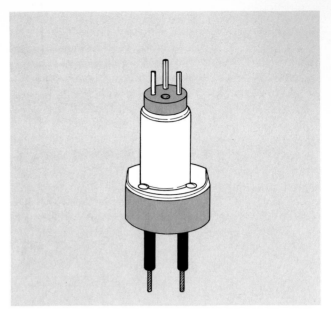

FIGURE 355. Low-voltage relay operates on 24-volts.

with each unit. Follow the manufacturer's instructions for installing the outdoor type.

Proceed as follows to install remote-control wiring:

1. *Install a 24-volt transformer (Figure 354).*

   Mount the transformer in a central location, since it will serve all the remote-control circuits. The transformer reduces voltage from 120 volts to 24 volts as required for the low-voltage system.

2. *Install a low-voltage relay at each lighting outlet box (Figure 355).*

   Mount a low-voltage relay in a knockout of each outlet box where a light is to be controlled. Connect the white wire from the source to the load (light). Connect the black wire from the source to the relay. Connect the other conductor from the relay to the load (light) inside the box. Three low-voltage wires extend outside the box for connecting to each switch.

3. *Run the low-voltage cable from the transformer to the relay and then to each switch position for that light.*

   On long spans, a messenger (support) cable may be required to support low-voltage cables from pole to buildings.

   Most of the cables used in low-voltage wiring consist of three parallel wires. Along one side of the cable is a ridge or a rib to identify the common conductor. **The common conductor (on the rib side) must be in the same position at each switch**

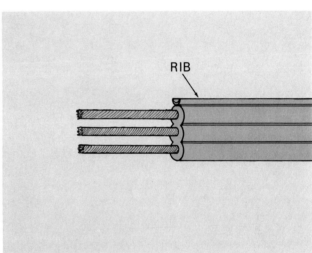

FIGURE 356. Common conductor identified by a rib on the top conductor of the cable must be connected to the same terminal position on every switch.

throughout the circuit (Figure 356). Some switches have numbered terminals making it easier to connect the common conductor to the correct terminal at each switch.

Allow about 15 cm (6 in.) of extra cable at each relay. The low-voltage cable is lightly insulated and does not require any further protection. It can be stapled on the surface or run through interior walls.

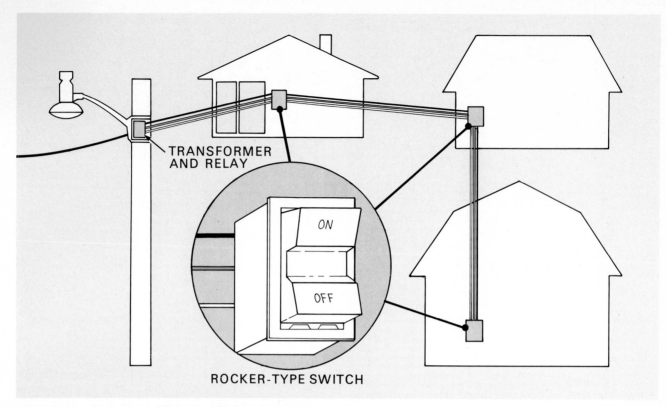

FIGURE 357 Typical yard-light installation of remote-control system.

4. *Install the switches (Figure 357).*

You may install as many switches as desired on any circuit. One type has a rocker-type switch which turns the light on when pressed at the top and off at the bottom. Others are the momentary-contact type that push to turn on and push again to turn off. These control a ratchet-type relay.

5. *Connect the switches.*

Connect wiring from the transformer to the relay and to each switch. Follow the manufacturer's instructions. Figure 357 illustrates a typical yard light installation. More switches could be added if needed.

F. Installing New Cable and Boxes in Old Buildings

In old buildings or homes, you may have problems extending circuits to new receptacles or fixtures. The following procedures will make the job easier. Turn off power to circuits at SEP before you start. For step-by-step procedures for connecting circuits, see pp. 78–91.

NOTE: If you are replacing a receptacle in a box that does not have a means of grounding, you must install either a non-grounding type or a ground fault circuit interrupter.

A new requirement in the 1993 *NEC*® mandates that where a receptacle is replaced and the current code requires a GFCI protected receptacle then the receptacle outlet must be upgraded to meet current code. *NEC*® 210-7(d).

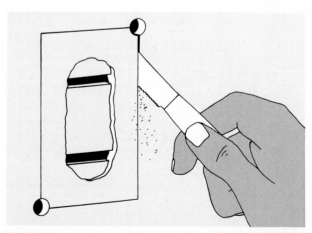

FIGURE 358. Use a template or outline of your box where the box is to be mounted. Drill holes at opposite corners and cut opening with a keyhole saw or hacksaw blade.

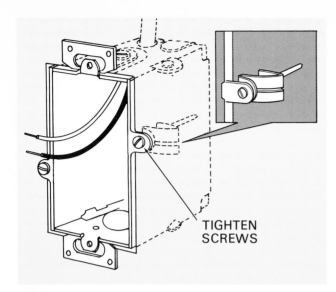

FIGURE 359. Install cable in box and fasten box in place with special expansion brackets and screws on each side.

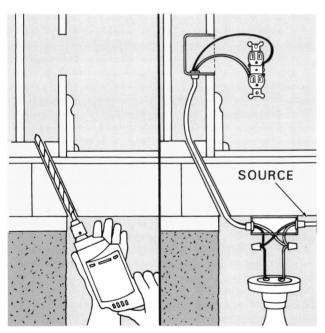

FIGURE 360. In basement, crawl space or first-floor closet in two-story home, use a long-shank bit to bore a diagonal hole through floor to new outlet opening. Use fish-tape to pull cable from closet fixture to new receptacle above.

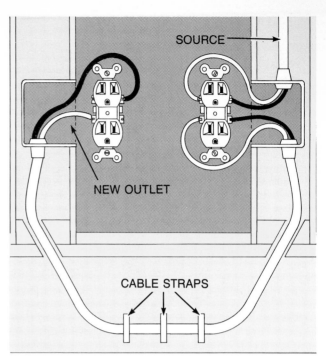

FIGURE 361. In basement or crawl space, run cable from old receptacle to new by boring two diagonal holes (or straight up on interior wall) to reach new opening. Staple cable to joist every three feet.

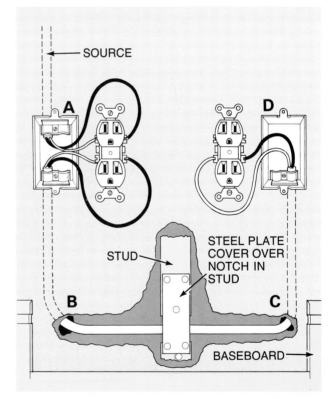

FIGURE 362. To run cable from old receptacle to new in the same room, cut opening D for new box. Remove the baseboard. Cut holes B and C and notch channel in plaster and stud. Fish cable from A to B to C and D. Install cable in new box and fasten in place. Cover notch in stud with steel plate and replace baseboard.

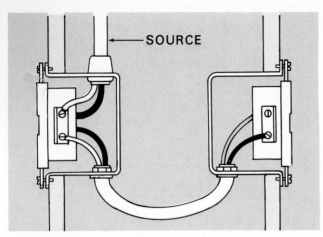

FIGURE 363. For back-to-back outlets, cut hole for new box opposite the old one in the next room, but offset slightly to allow room for the box. Run cable from old receptacle to new box. Clamp cable to new box and install the box in the opening.

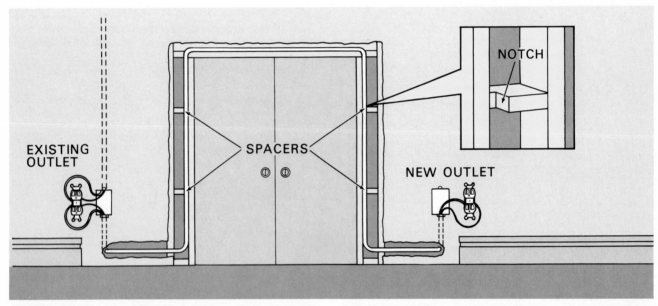

FIGURE 364. To run cable around a door frame, remove baseboards and door trim. Notch the wall plaster and the wood spacers between the frame and jamb. Run cable from old receptacle around door frame and to new box. Clamp cable to new box, install the box in the opening and fasten in place. Cover notches with steel plate.

Acknowledgments

The following assisted with the first edition of this publication. Their status and location were current at the time of the initial printing.

INDIVIDUALS

Anderson, Paul M., Agricultural Engineering Department, Pennsylvania State University

Barr, Robert C., National Fire Protection Association, Boston, MA

Bear, W. Forrest, Agricultural Engineering Department, University of Minnesota

Bennett, Wayland, Agricultural Engineering Department, Texas Tech University

Blanton, Lloyd H., Agricultural Education Department, Clemson University

Braker, Clifton R., Agricultural Engineering Department, University of Arkansas

Brown, Robert H., Chairman, Agricultural Engineering Division, University of Georgia

Bruwelheide, Ken, Agricultural and Industral Education, Montana State University

Bryan, C. E., Reynolds Aluminum, Richmond, VA

Bullard, Ben B., Texas Farm Bureau, Waco, TX

Carlson, Axel R., Extension Engineer, University of Alaska

Carter, James K., Jr., Agricultural Mechanics Department, Reedley College

Clouse, James P., Agricultural Education Department, VPI & SU

Collins, Edmond B., Cooperative Extension Service, West Virginia University

Cranford, William, Central Power and Light Co., Corpus Christi, TX

Crawford, Fred M., Extension Agricultural Engineer, University of Missouri

Crawford, Ronald, Washington State Department of Education

Cross, Ward H., Texas Farm Bureau Insurance Companies, Waco, TX

Daum, Donald R., Agricultural Engineering Extension, Pennsylvania State University

Davies, Roger, Raco, Inc., South Bend, IN

Davis, John R., Agricultural Engineering Department, Oregon State University

Deming, D. R., Raco, Inc., South Bend, IN

Doering, Floyd J., Wisconsin State Department of Education

Downey, Gary, Grinell Mutual Reinsurance Co., Grinell, IA

Durham, Dwight H., Southwire Co., Carrollton, GA

Durkee, James, Department of Vocational Education, University of Wyoming

Espenschied, Roland F., Agricultural Engineering Department, University of Illinois

Farmer, Gary, Georgia State Department of Education

Faulkner, T. L., Alabama State Department of Education

Fiester, Fred, Claverack R. E. C., Towanda, PA

Froehlich, Nord, Square D Company, Lexington, KY

Gerrish, John, Agricultural Engineering Department, Michigan State University

Giguere, John F., Sault Ste. Marie, Ontario

Gilbertson, Ozzie, Agricultural Education Department, University of Nebraska

Gilden, Robert O., Agricultural Engineer, ES-USDA, Washington, DC

Gilman, Francis E., Extension Agricultural Engineer, University of New Hampshire

Ginn, Olin W., Georgia Power Company, Atlanta, GA

Grant, Lee, Agricultural Engineering Department, University of Maryland

Griffith, Lowell, Illinois Power Company, Decatur, IL

Hansen, Herbert, Agricultural Education Department, Oregon State University

Hansen, Hugh J., Western Regional Agricultural Engineering Service, Oregon State University

Harvey, L. M., Electrical/Electronic Department, The Sault College of Applied Arts and Technology

Hayles, Jasper A., Agricultural Education Department, Arkansas State University

Hinkle, Charles N., Agricultural Engineering Department, Purdue University

Hoerner, Thomas A., Agricultural Engineering Department, Iowa State University

Hogsett, Ordie L., Extension Safety Specialist, University of Illinois

Holler, Ivan, Wayne-White Cooperative, Fairfield, IL

Huff, Edward R., Agricultural Engineering Department, University of Maine

Jacobs, Clinton O., Agricultural Education Department, University of Arizona

Killough, Frank B., Alabama State Department of Education

Kimmons, Marion L., Agricultural Mechanics Department, West Virginia University

183

Kindschy, Dwight, University of Idaho

Kingsley, Curtis, Georgia State Department of Education

Kruesi, William R., General Electric Company, Fairfield, CT

Land, Bill, Georgia Power Company, Atlanta, GA

Law, S. Edward, Agricultural Engineering Department, University of Georgia

Lawrence, John A., Agricultural Education Department, University of Idaho

Lively, Robert M., Gifford-Hill & Company, Inc., Lubbock, TX

Matt, Stephen R., Industrial Arts Department, University of Georgia

McBride, Robert, Vo-Ag Teacher, Kenton, OH

McClure, W. D., Agricultural Engineering Department, Texas A&M University

McFate, Kenneth L., Food and Energy Council, Inc., Columbia, MO

McLendon, Derrell, Agricultural Engineering Department, University of Georgia

Meeks, Leon, Electrical Construction, North Georgia Technical School

Miller, Bruce W., Central Power and Light Company, Corpus Christi, TX

Mitchell, Kenneth K., Tennessee State Department of Education

Mitchell, Martin L., New Hampshire State Department of Education

Moore, Carlos, Arizona State Department of Education

Moore, James, Electrical Inspector, Decatur, IL

Moore, Milo J., Extension Engineer Specialist, University of Vermont

Moss, Carter, Georgia Power Company, Atlanta, GA

Mumford, Michael, Agricultural Engineering Department, University of Nebraska

Munson, Maynard, Iowa Power Company, Des Moines, IA

Nelson, Robert, Mohawk College of Applied Arts and Technology, Hamilton, Ontario

O'Brien, Michael, Agricultural Engineering Department, University of California-Davis

Page, Foy, Vocational Instructional Services, Texas A&M University

Parrish, Bob, Residential Wiring, DeKalb Occupational Education Center

Patton, Bob, Oklahoma State Department of Vocational Technical Education

Pearson, A. J., Asst. Dir., National Joint Apprenticeship and Training Committee, Lanham, MD

Phillips, Andrew, Director, National Joint Apprenticeship and Training Committee, Lanham, MD

Plummer, Robert K., Maine State Department of Education and Cultural Affairs

Priebe, Don, Agricultural Education Department, North Dakota State University

Reed, Joe, Agricultural Engineering Department, University of Georgia

Rice, Charles, Agricultural Engineering Department, University of Georgia

Ridenour, Harlan E., Ohio Agricultural Education Curriculum Materials Service

Searls, Dean, Adams Electric Cooperative, Camp Point, IL

Sheets, J. T., Electrical Inspector, Decatur, GA

Slyter, Damon E., Kansas State Department of Education

Smith, Kenneth L., A. O. Smith Harvestore Products, Inc., Eureka, IL

Smith, Norman, Agricultural Engineering Department, University of Maine

Stetson, LaVerne E., Agriculture Engineering Farm Electric Research, University of Nebraska

Surbrook, Truman, Agricultural Engineering Department, Michigan State University

Tart, C. V., North Carolina State Department of Public Instruction

Todd, John D., Agricultural Education Department, University of Tennessee

Trollope, D. A., Apprentice Department, Mohawk College

Warnock, William K., Agricultural Engineering Department, University of Arkansas

Watts, Thomas, Vermont State Department of Education

West, William, West Virginia State Department of Education

Wills, William D., Agricultural Engineering Department, California Polytechnic State University

Wilson, Donald E., California State Department of Education

Winsett, Ivan, Ronk Industries, Nokomis, IL

Wood, Jay, Washington State Department of Education

INDUSTRY

A.O. Smith Harvestore Products, Inc., Eureka, IL

Black and Decker, Towson, MD

Blackhawk Industries, Dubuque, IA

Bush Grip

Bussman Manufacturing, Earth City, MO

Carlson Electrical Services, Inc., Cleveland, OH

Challenger Electrical Equipment Corp., Malvern, PA

Channellock, Inc., Meadville, PA

Crouse-Hinds, Syracuse, NY

Farm Electric, Inc., Athens, GA

Georgia Power Co., Athens, GA

General Electric Company, Fairfield, CT

Gifford-Hill & Company, Inc., Lubbock, TX

Gould ITE, Springhouse, PA

Honeywell, Minneapolis, MN

Hubbell, Harvey, Inc., Christianburg, VA

Ideal Industries, Sycamore, IL

I.T.T. Holub, Sycamore, IL

Klien Tolls, Inc., Chicago, IL

Larkin, G.E. Company, Hickory, NC

Levitron, Little Neck, NY

McGill Manufacturing Co., Inc., Valparaiso, IN

National Fire Protection Association, Boston, MA

Raco, South Bend, IN

Reynolds Aluminum, Richmond, VA

Slater Electric, Inc., Glen Cave, NY

Southwire Company, Carrollton, GA

Square D Company, Lexington, KY

Texas Farm Bureau Insurance Company, Waco, TX

Thomas and Betts, Inc., Bridgewater, NJ

3M Company, St. Paul, MN

Wiremold Company, West Hartford, CT

Index